T0258936

ECOLOGICAL RISK EVALUATION OF
POLLUTED SOILS

Ecological Risk Evaluation of Polluted Soils

Jean-Louis Rivière
Unité de Phytopharmacie et Médiateurs Chimiques
Institut National de la Recherche Agronomique (INRA)
Versailles
France

A.A. BALKEMA/ROTTERDAM/BROOKFIELD/2000

Aidé par le ministère français chargé de la culture.
Published with the support of the French Ministry of Culture.

Translation of: *Évaluation du risque écologique des sols pollués* 1998,
© Technique & Documentation, Paris.

Translation editor : Latha Anantharaman
Technical editor   : Dr. Kaiser Jamil

Authorization to photocopy items for internal or personal use, or the internal or personal use of specific clients, is granted by A.A. Balkema, Rotterdam, provided that the base fee of US$1.50 per copy, plus US$0.10 per page is paid directly to Copyright Clearance Centre, 222 Rosewood Drive, Danvers, MA 01923, USA. For those organizations that have been granted a photocopy license by CCC, a separate system of payment has been arranged. The fee code for users of the Transactional Reporting Service is: 90 5410 796 0/2000 US$1.50 + US$0.10.

A.A. Balkema, P.O. Box 1675, 3000 BR Rotterdam, Netherlands
Fax: +31.10.4135947; E-mail: balkema@balkema.nl

Distributed in USA and Canada by
A.A. Balkema Publishers, Old Post Road, Brookfield, VT 05036-9704,
USA Fax: 802.276.3837; E-mail: Info@ashgate.com

ISBN 90 5410 796 0

© 2000 Copyright Reserved

Preface

The attention given to environmental pollution is very recent—not dating more than two generations. Initially, a new scientific field, ecotoxicology, was evolved to study the impact of pollutants on natural ecosystems, but very soon, people were no longer content to accept pollution as an inevitable companion to technical progress, and they felt the need to limit the dispersal of growing quantities of chemical products into the environment.

Custodians of the environment were thus confronted with the need to take decisions—particularly regulatory decisions—on the most rational basis possible, and a new interdisciplinary field emerged, the evaluation of the ecological risk posed by pollutants.

The subject is vast and complex. Our understanding of natural ecosystems is still very imperfect, and prediction is by nature an uncertain operation, but a look at the scientific literature shows that research is very active, and that methods and techniques are rapidly evolving.

This book is an introduction in which the reader will find the fundamental principles, as they are conceived at present, and a range of workable methods. I have tried to ensure a treatment of the subject as a whole, but, inevitably, some parts are more developed than others, for different reasons: lack of competent personnel, abundant treatment of an issue in other works, or lack of scientific knowledge in a particular field. The last chapter, for example, is dedicated to risk formulation and management, a frontier area. The subject cannot be passed over without comment, but it merits, obviously, a larger development.

I owe a great deal to the support of Alain Navarro, professor at INSA (Lyon), and Scientific Director of RECORD, who from the beginning showed a lively interest in the subject and constantly encouraged me during the editing, and also to RECORD, which financially supported the publication of this work. I also wish to thank all my colleagues, family, and friends, who have enabled me, in their various capacities, to bring this work to fruition.

Jean-Louis Rivière
Versailles, 2 May, 1997

Contents

Introduction:
Soil Pollution and Ecotoxic Risk

The idea that the earth is a closed system and that soil, like other mediums, is polluted by human activities, is very recent, hardly thirty years old. The chief preoccupation has been with water pollution, a conviction that, sooner or later, all the pollutants found in water were the principal cause of the emergence of aquatic ecotoxicology. Yet, the existence of polluted soils has been cited since ancient times. Greek and Roman writers remarked that the contamination of water and air near mines had adverse effects on plants, domestic animals, and humans. But soil pollution is not as visible as water pollution, and to acknowledge that soils can be polluted goes against the belief—still very widespread—that they have an unlimited capacity to purify themselves. Perceptions have evolved: DDT pollution,[1] the Seveso catastrophe (1976), urban pollution by pyralene electric transformers (Reims, 1985; Villeurbanne, 1986), and the nuclear fallout at Chernobyl (1986) have clearly shown that environmental pollution is general and that it affects soil as well as other mediums. Ancient practices, such as the spreading of purifying mud around farming areas, earlier considered a wasteful agricultural amendment, are now being considered again. There is now an increasingly strong public demand for clean soil, as for clean water and clean air, to ensure the well-being of humans and other living species.

The quality of soils is of great importance, as emphasized in the report of INSA/INRA/CRIDEAU/CNRS (I2C2, 1994):

- Soil is a living medium much more complex than air or water. It plays an essential role in the production of biomass and in the recycling of elements, and its functional characteristics can be altered by pollutants.

---

[1]DDT: 2,2-*bis*-(*p*-chlorophenyl)-1,1,1-trichloroethane.

- Soil pollution can affect other mediums and plants, and can ultimately reach terrestrial and aquatic animal species.

The authors of I2C2 (1994) distinguished between two types of pollution:

- *Diffuse pollution*, affecting large land surfaces, resulting from the dispersion, probably by atmospheric means, of phytosanitary products and industrial pollutants. These situations lead to *polluted soils*.
- *Localized pollution*, much more intensive, resulting from the spilling, accidental or otherwise, of solid or liquid products, leads to *polluted sites*.

To these spatial criteria may be added some temporal criteria:

- Sites that *have been* polluted because of old mines, industrial contaminations, abandoned discharges.
- Sites that *are being* polluted by industrial, agricultural, or domestic activities. To this pollution caused by human activity is added *natural pollution* of the environment, for example the existence of significant geochemical beds of metals.
- Sites that *will be* polluted[2] by the presence of new chemical products, or by new industrial or agricultural activities.

Whatever the origin and duration of the pollution, the problem scientists must solve is to define criteria of quality that allow soils to be classified as 'clean' or 'polluted'.[3] In some countries, the concept of 'multifunctionality' of soil would be used (the soil is clean for all purposes) and, in others, standards of quality (maximum concentrations of pollutants) would be established according to the uses of the soil; a high quality is demanded for agricultural lands, because of the risk of food contamination, or for ecologically very sensitive mediums (e.g., natural reserves). The determination of quality criteria ultimately has two consequences:

- the obligation to rehabilitate very polluted soils to bring the concentrations of pollutants to acceptable levels;
- the obligation to prevent and control future pollutions for which the threshold of danger is not yet crossed.

The first operation is very costly. The determination of threshold thus clearly has economic consequences for the community. In the second case, regulatory constraints bear more and more forcefully on manufacturing industries, which must take them into account in their operation, even if they do not control all the stages of their production cycle (production,

---

[2]Here there is a risk of pollution.
[3]This definition is essential: apart from some extreme situations of pollution that are well defined (certain industrial sites), as long as there are no established criteria, one cannot say of a site or soil that it is polluted. In many cases, the pollution of a site is suspected, and must be then confirmed or disproved.

transport, use, and disposal). The development can be very rapid: the example of the fertilizer industry in the course of the past thirty years shows the level of information that is now necessary before allowing into the environment a product that presents *a priori* certain environmental risks. Waste management has become a worrying problem, because uncontrolled waste that pollutes water and soil is more and more severely regulated.

The various strategies developed to evaluate the quality of soils and sites correspond to three possible objectives:

- to establish *references* or criteria of soil quality, on chemical and/or ecotoxicological bases (to define thresholds);
- to develop *methods of ranking* to classify polluted sites for the purpose of their decontamination (to establish a classification); and
- to develop *methods of risk evaluation*, comprehensive or simplified, to define the ecotoxic impact (to measure a risk).

These different approaches have been used with varying success, as shown by the difficulties found with ranking systems such as the American Hazard Ranking System. They correspond, in effect, to risk evaluations based on more or less detailed data, and based on the implicit hypothesis that the greater the concentration of the pollutant, the greater is the risk of toxic effects. In theory, these methods serve to resolve different problems: the use of references seems appealing to define the quality of a polluted soil and fix a limit above which it can no longer be adapted to certain uses. A ranking method is certainly useful to classify sites and determine decontamination priorities, whereas risk evaluation strategies, more difficult to carry out, could be reserved for populations or ecosystems particularly at risk.

Ecotoxic risk comprises *health risk* (risk to the health of human populations) and *ecological risk* (risk to the environment[4]). The general principles of the evaluation of these two types of risk are presented in chapter 1 of this work.

Strategies for the evaluation of ecological risk aim to define as completely as possible the potential or inconspicuous effects of pollutants on the environment, and the chief objective of the evaluation will be to designate the *elements at risk* (animal or plant populations, functional capacity

---

[4]It is not easy to define what exactly is understood by environment. For example, in directive 91/414 of the European Union, concerning risk evaluation for phytopharmaceutical products before they are put on the market, the environment is defined as 'the water, air, land, wild fauna and flora, as well as all the interrelations between these various elements and all relations existing between them and every living organism.' This definition is interesting, because it mentions not only the various constituents of an ecosystem, that is, the mediums and the animal and plant populations, but also relations between these constituents, which is to say that the final objective of evaluation is actually the ecosystem even if this term is not used.

of an ecosystem, etc.), in close interaction with decision-makers, those responsible for polluted sites, the public, and others. This initial stage of *risk formulation* will be elaborated in chapter 5, with the final stage of *risk management*. This arrangement is justified if one considers that risk evaluation is initiated by the manager, and that it is the manager who will make use of its results.

Once the elements at risk are identified, the existing scientific data can be used to work toward evaluating the modalities and extent of contact between the elements at risk and the pollutant (*characterization of exposure*; chapter 2), in parallel with an evaluation of the relation between the dose and the effects (toxicity) of the pollutant (*characterization of effects*; chapter 3). Finally, the risk is characterized by an evaluation of the extent of predicted effects and of the probability of their realization, as a function of exposure (chapter 4). The necessary data are obtained by various approaches: the occurrence and behaviour of products in air, water, and soil are characterized by laboratory assays, measurements made on the land or simulated by mathematical models; the estimation of toxic effects of pollutants is based on the same methods, laboratory studies on different plant or animal species, epidemiological studies of plant, animal, or human populations, or mathematical models.

Definitions, Concepts and General Models of Risk Evaluation

1. DEFINITIONS

1.1 Danger, Risk and Risk Evaluation

There are several definitions of risk evaluation, and they enable us to specify the nature and the impact of this operation. Risk evaluation is 'an operation that assigns levels and probabilities to adverse effects[1] of human activities and natural catastrophes'[2] (Suter, 1993a). The evaluation of health risk has been defined as 'the evaluation of information about the intrinsic danger of substances, the degree of human exposure to such substances, and description of the risk that arises from that exposure' (National Research Council, 1994). For Covello and Merkhofer (1993), risk is a concept 'at least two-dimensional, implying (a) the possibility of an adverse effect and (b) an uncertainty about the appearance, chronology, and gravity of this adverse effect. If one of these characteristics does not exist, there is no risk.... More formally, risk [is] the characterization of a situation or action from which two results are possible, and it is not known which will occur, and one of them represents an undesirable event.' According to the same authors, risk evaluation is 'a systematic operation to describe and quantify the risks associated with dangerous products, operations, actions or events.' Other definitions have been proposed. For example,

---

[1]An adverse effect is an alteration of the functioning capacity or the capacity to recover normal function within a reasonable interval after a stress; for example, an animal that has an immune deficiency is incapable of resisting a microbial attack, which renders it very vulnerable. To define an adverse effect is relatively easy in health risks or for ontoecological populations, but it is much more difficult for an ecosystem (cf. chapter 5).

[2]We will later come across this idea that risk evaluation is not confined to chemical products but applies to all forms of stress.

Volmer et al. (1988) define risk evaluation as 'methods designed to estimate the significance and probability of adverse effects of anthropogenic substances on the environment.' These various definitions do not always specifically refer to a particular type of risk, health or ecological. According to Norton et al. (1992), the evaluation of ecological risk is 'an operation that evaluates the likelihood of adverse ecological effects produced as a result of exposure to stresses.'[3]

The most recent definition is that of Rodricks (1994): 'risk evaluation ... is a systematic means of organizing available information and knowledge and specifying the level of scientific certainty, in relation to the facts, models, and necessary hypotheses; the objective is to draw conclusions from these about health risks, of whatever nature.' This definition is very interesting because it brings to light the essential elements of the operation of risk evaluation:

- research and organization of existing information;
- use of different approaches and methods;
- specification of an uncertainty attached to a result.

The necessity of making a risk evaluation, even summarily, lies in a double observation:

- One cannot eliminate the possibility of unpredictable adverse effects of human activity (one cannot foresee everything).
- Some decisions must be taken, even on the basis of necessarily incomplete information (one cannot wait).

Risk evaluation does not apply only to chemical products. Insurers who must predict the risk of hail, floods, or automobile accidents to calculate the schedule of premiums, and engineers who must calculate the probability of rupture of a canal or the rate of rejection of defective pieces practise particular forms of risk evaluation.

Risk evaluation is founded on the fundamental distinction between *danger* and *risk*. In the case of chemical products,[4] the danger is linked to the existence of dangerous substances, that is, those that have the *potential to exercise adverse effects on the environment and living species, if they come into contact with them*. Contrary to what is often asserted, it is not properly speaking an intrinsic property of molecules (as would be molecular weight or the number of carbon atoms, for example), since the danger is relative to the living element or medium in which the molecule acts. One can allow this point of view in health risks, since the danger is always measured in relation to a single biological system—the human—but it is not justified in ecological risks: all species are not equally sensitive to toxins.

---

[3]For the definition of *stress*, see the end of this section.
[4]The regulatory terminology earlier used the term *substance*.

Dangerous products are distinguished from others by their capacity to cause toxic effects in the short term (mortality) or in the long term (occurrence of cancers, reproductive problems, etc.). Moreover, this definition must be accompanied by a notion of *dose*. Certain chemical products necessary to the proper functioning of the organism may prove to be toxic in very high doses, or, on the contrary, in very low doses, they can be the origin of deficiencies the expression of which can be considered a particular form of toxicity. The classic examples of fluoride and selenium show that the notion of dangerous product falls within sometimes very narrow limits. The *danger* arises from the substance itself or from the substance and environmental components that are closely mixed with it (matrix). The fumes of incinerators, the mud from waste treatment plants industrial effluents, automobile emissions, and a badly polluted medium (for example, the soil in a site containing significant quantities of potentially toxic pollutants) are *dangerous objects*.

The risk is the *probability of occurrence of toxic effects after exposure of the organism to a dangerous object*. The notion of risk takes into account the existence of a possible exposure to dangerous objects. The risks may be larger with a less dangerous pollutant spread out over large areas over long periods of time, than with a very dangerous product produced in small quantities, stored under good conditions, and reserved for very limited uses.

It is important to distinguish between *pollutant* and *toxin*: a very dangerous product kept confined in a laboratory, in small quantities, is a toxin, but not a pollutant. Conversely, a pollutant is not always a very toxic product, but the capacity of a chemical substance to disperse through the environment in large quantities classes it automatically as a pollutant, that is, a product presenting a potential risk for that environment. It is this that the European Union implicitly recognized when it demanded a large number of ecotoxicity tests when the quantity of a dangerous substance produced rises, in direct proportion to the probability of dispersal in the environment. For Ramade (1993), a toxin is 'a substance that can be absorbed ... and that causes an intoxication of organisms affected that can lead to death' and a pollutant is 'a substance (or a process) of physical, chemical, or biological nature, capable of contaminating various types of terrestrial, limnic, and marine ecosystems'. For Moriarty (1988), 'a pollutant is a substance that is found in the environment, at least partly as a result of human activities, and that has a harmful effect on living organisms'. In the majority of texts, *contaminant* and *pollutant* are used with the same meaning, but for Ramade (1933), a contaminant is 'a pollutant present in detectable quantities in the environment or in an organism'.

Other terms are used. For example, Sipes and Gandolfi (1991) define *xenobiotics* as substances foreign to the organism or that do not seem—to our present knowledge—to be indispensable to the normal functioning of the organism. This term is less restrictive than *pollutant* or *toxin*. It does not presuppose the possibility of adverse effect and encompasses not only anthropogenic pollutants, but also a number of natural substances present in plants and found in the food of humans and animals, such as flavones, terpenes, etc., which are just beginning to be suspected of biological effects. Rand and Petrocelli (1985) define a *xenobiotic* as a 'foreign constituent that is not produced by nature and that is not normally considered a constitutive part of a particular biological system' and apply this term to products of industrial origin. Their definition has the disadvantage of excluding the natural pollutants, such as heavy metals, and elements such as fluoride or selenium, indispensable to the normal functioning of the organism, but toxic in large doses.

In the text that follows, *pollutant* is defined as *a dangerous object capable of presenting a risk to environments and living organisms*. A *polluted site* is a *geographic zone in which pollutants are found*. *Pollution* is defined as *the actual or supposed presence of pollutants in the environment*. The terms *pollutant* and *contaminant* are synonymous most of the time, and in the following text, we use the two interchangeably.

Chemical substances are not the only environmental dangers: climatic changes, modifications of rural areas, etc., are threats to existing ecosystems. The present trend is to group all these potential dangers under the general term of *stresses*. In the same manner, individuals, environments, or ecosystems susceptible to stressful effects are designated under the general term of *elements at risk* or *receptors*. We will see later that models of ecological risk evaluation must frequently use general terms that enable us to show the common bases of diverse strategies of risk evaluation and to assess the environmental risk of chemical products in relation to other factors that disturb ecosystems.

1.2. Ecotoxicology and Risk Evaluation

All authors now agree in attributing the origin of the word *ecotoxicology* to Truhaut (1977), but it is difficult to know what exactly is included in the term to distinguish it from environmental toxicology, how ecotoxicology is defined in relation to ecology and toxicology, and what the particular place of risk evaluation is in relation to these different disciplines.

In the absence of universal agreement, *ecotoxicology* is defined here as the *study of the occurrence of pollutants and their effects on the environment and humans, that is, abiotic mediums and the biotic components that populate them*. This definition is very wide, since it includes the occurrence and effects of pollutants under the same term; also, it takes into account the direct effects of pollutants on living organisms and the direct effects on environments (for example, the greenhouse effect on the ozone layer) and the indirect repercussions on biocenoses.

This definition does not specify the level of organization of biological system: one of the characteristics of ecotoxicology often emphasized is to consider ecosystems and not just individuals, but sometimes a 'toxicology' of the individual has been opposed—wrongly—to an 'eco' toxicology that takes only ecosystems into account. According to Barbault (1993), 'as a basic science, ecology has as its objective the study of the organization, functioning, and evolution of biological systems corresponding at an equal or higher level of integration to that of the individual.' The definition proposed by Barbault is very wide, since it takes into account not only the level of ecosystems, but also that of communities, of populations (biology of populations), and of individuals (ecophysiology). He insists, moreover, on the central role of the individual, which is the object immediately 'given' and visible to the observer. Suter (1993a) makes the same remark: the individual is a biological system immediately perceptible to the observer for two essential reasons:

- Ecotoxicology is derived from human toxicology.
- The individual is the most legible entity, the most easily observed and interpreted in an immediate perception. The lower levels (cellular and molecular) are not visible except with instrumentation that presupposes a learning and deformation of reality; the higher levels (communities and ecosystems) are also not understood without some degree of theoretic conceptualization.

The study of individuals is indispensable (cf. chapter 3) to the prediction of ecological risks, because although it is possible to measure the disastrous effects of certain pollutants on ecosystems after the fact, it is not possible to predict these effects simply by direct studies of the ecosystems. There is, therefore, no reason to limit the field of ecotoxicology exclusively to the study of ecosystems, even though they constitute a research objective that must be a priority.

The different applications of ecology have been pointed out by Barbault (1993): regulations of pest or exploited populations, preservation and use of genetic diversity, agricultural practices (biocontrol, for example), management of territory, and conservation of fauna and flora. The principal application of ecotoxicology is *the evaluation of the risk posed by chemical products to the environment and to humans*.

The difference between ecotoxicology and risk evaluation is important. A very eloquent analogy can be found in the example of climate: on the one hand, a fundamental science, climatology, enables us to understand and explain climatic phenomena, and on the other, an applied science, meteorology, provides the climatic predictions necessary for human activities. The relations between ecotoxicology and risk evaluation are, in two senses, like those between climatology and meteorology. Ecotoxicology provides the scientific bases and facts that enable risk evaluation, but conversely, the needs of risk evaluation create and generate ecotoxicological studies. In the same way, the need for meteorological predictions sustains climatological studies, which, in turn, constitute the scientific bases of these predictions.

1.3. Development of Risk Evaluation Strategies

Risk evaluation arose when people recognized that they use toxic products for their vital needs. It became necessary to manage the use and handling of such products, at first informally, through advice, advertisements, and recommendations, and then by the more stringent means of regulations and sanctions. These regulations operate on empirical bases, rarely spelled out, which lead to decisions that are not always scientifically justified, being too prudent for some and insufficient for others. The need to develop a rational strategy for decision making (regulatory or relating to regulation) and define environmental management practices *from existing scientific data* has led to the rise of risk evaluation as a scientific discipline, with its own vocabulary and methods.[5]

The methods of risk evaluation were developed principally in the United States, in order to satisfy the requirements of numerous laws promulgated in the 1970s and 1980s, which reflect the environmental preoccupations of that country (for example, CERCLA, FIFRA, SARA, and TSCA[6]). In the 1980s, several commissions formed out of the National Academy of Sciences drew up methodological bases of risk evaluation: these are now used by numerous federal agencies such as the EPA or FDA.[7]

A specialized commission (the Risk Assessment and Management Commission) was created with the mission of evaluating the current standards and methods of risk evaluation and making recommendations on the best use of available information (*Scientist*, 1994). In the same spirit, some more or less detailed procedures were elaborated in other countries.

The evaluation of ecotoxic risks is a recent and complex scientific field, with a significant conceptual base that does not have a fixed and unanimously accepted vocabulary.

---

[5]The reader will find a comparative analysis of modes of operation in the 1950s and today, as well as recommendations for the future, in an article by the renowned toxicologist John Doull (Doull, 1996).

[6]CERCLA, Comprehensive Environmental Response Compensation and Liability Act; FIFRA, Federal Insecticide, Fungicide and Rodenticide Act; SARA, Superfund Amendment and Reauthorization Act; TSCA, Toxic Substance Control Act.

[7]EPA, Environmental Protection Agency; FDA, Food and Drug Administration.

1.4. From Human Risk to Ecological Risk

We have earlier seen that evaluation of ecotoxic risk can be subdivided into two principal branches, the evaluation of *risk to human health* (health risk) and the evaluation of *ecological risk* (risk to the physical environment and plant and animal organisms other than human). This distinction reflects a historic shift: evaluation of risks associated with chemical products was originally designed to protect human health. Despite the work of pioneers such as Cairns and his colleagues, who, since 1978, defined 'the evaluation of [ecological] risks as an operation whose objective is to provide information supporting a judgement of the degree of toxicity of a substance or of an activity', it is only recently that the concepts and methods of health risk evaluation have been extended to the evaluation of risk to ecosystems, chiefly since the works of Barnthouse and Suter at the Oak Ridge National Laboratory (USA) in the 1980s. The reasons for this delay are linked to a slow awareness of the potential risk to the environment, and also to the fairly widespread belief that measures taken to protect human health ensured a high degree of protection, sufficient to ensure the health of other organisms. Some examples show the poverty of this argument (Suter, 1993a):

- DDT and its metabolites have had adverse effects on some bird populations, without parallel with the effects so far observed in humans.
- PCDD/PCDF[8] are much more toxic to several animal species than to humans: the pollution of the Love Canal (USA) had pronounced toxic effects on rodent populations (sterility and precocious mortality) and some bird populations suffered from the pollution of the Great Lakes (embryo mortality and teratogenesis).

It is now acknowledged that ecological receptors are sometimes more sensitive than humans (Bascietto et al., 1990). The evaluation of health risk and the evaluation of risk to other animal species are based on identical principles, but it was quickly recognized that the diagrams that were developed in the first case are not well adapted to the second. According to Suter (1993a), the divergences occur on the following points:

- Animals are exposed by avenues that are unique to them, for example, the grooming of fur in small mammals.
- Given the very large number of animal species, the probability of finding one or several species more sensitive than humans is mathematically not negligible. The cause of these interspecific differences is not always known. In some cases, it is the consequence of anatomic

---

[8]PCDD, polychlorodibenzo(p)dioxins; PCDF, polychlorodibenzofurans.

or functional characteristics that do not exist in humans, for example the thinning of the shell of bird eggs caused by DDT.

- The large-scale phenomena of the ecosystem do not have a human equivalent, for example, the eutrophication of a lake or its acidification by acid rain.
- Species other than humans are subject to stronger exposure, for example, because of monophagous diets (a heron consumes only fish, while a human has a varied diet) or because of closer contact with the ambient medium (immersion in water for fish, close contact with the earth for small mammals and earthworms).[9]
- Most birds and mammals are smaller than humans, and their energy metabolism more intense, which means that these species consume more contaminated food, drink more contaminated water, and breath larger volumes of polluted air (in relation to their unit of mass).
- Certain products are specially designed to fight pest species and inherently present a significant risk to neighbouring species on the phyllogenic plain (a herbicide presents higher risk to plants than a neurotoxic insecticide). The risk can also come from spatial coincidence between the pollutant and the polluted, for example, the individuals present in a field during a pesticide treatment are more exposed than individuals further away.
- Animal species are more closely allied to their environment than humans, who can always, at least theoretically, avoid certain dangers by varying their diet, eliminating certain foods, or changing their domicile.

The different points of divergence between human risk and ecological risk pointed out by Suter do not all have the same weight. For example, the existence of different levels or avenues of exposure does not justify different strategies in risk evaluation. Furthermore, health risk is evaluated for individuals, but the impact of pollution is studied and evaluated on different geographic scales, proximity to a site, city, or country, or on a continental scale, which implies the consideration of human populations.

For Lipton et al. (1993), the ecological risk differs from human risk on four essential points:

- The identity of receptors is unknown. The evaluation of human risk, since the beginning, has been focussed on the human, while the elements at risk are much more difficult to define in an evaluation of ecological risk. For example, the effects of DDT on invertebrates

---

[9]According to Burger and Gochfield (1992), the exposure is more difficult to evaluate in an ecological risk than in a health risk, because of the varying life spans of the organisms that make up a trophic network: the time scales are not the same.

is predictable, since the product is a powerful insecticide, but not its effect on birds.

- The receptors are located at different levels of biological organization. Health risk considers individual humans, while the evaluation of ecological risk must include populations, ecosystems, and eventually ecocomplexes. The procedures developed for the evaluation of individual risk are rarely pertinent to the understanding of phenomena at a higher or lower level of biological organization.
- The selection of final points[10] supposes the prior definition of identity of receptors and their level of organization, which is not necessary in health risk.
- The existence of retroactive effects. The adverse effects on a population are a source of risk to other elements of the ecosystem: one must take into account the possibility of cascading risks, for example, the disappearance of functionally important (or key) species in a community or ecosystem.

We can find an excellent example of retroactive effect, this time between causes and effects of pollutants, in the work of Forbes and Forbes (1994). Benthic species act on the dispersion of pollutions in the sediments (by their search mechanisms) and, reciprocally, the variations of exposure that result from that modify the toxicity of pollutants for the same species. This example is very interesting, because identical phenomena are found in soils, notably under the action of earthworms.

To summarize these different arguments, one can say that health risk evaluation differs from ecological risk evaluation by its *nature* and by the *number* of *final points*:

- The number of species: a single species in the case of health risk, millions of species in the case of ecological risk.
- The level of biological organization: health risk is concerned essentially with the risk for some individuals and populations at risk; the evaluation of ecological risk is supposed to encompass the effects at the higher levels of biological organization, communities and ecosystems.

1.5. Risk Evaluation and Related Disciplines

The methods of ecological risk evaluation are useful aids to decision making designed to take into account a certain number of environmental problems, not just chemical risk. They are used in parallel with other practices that can either help to develop these strategies or benefit from them in meeting their own objectives.

---

[10]For the definition of *final point*, see section 3.2.2.

1.5.1. Evaluation of Environmental Impact

According to Erickson (1994), the evaluation of environmental impact is 'an operation aiming to identify and evaluate the consequences of human activities on the environment and, if necessary, to remedy them'. It is a predictive approach that aims to establish the environmental consequences of operations of infrastructures, such as the building of factories, electric lines, highways, etc. A detailed description of the objectives and methods of this discipline can be found in the work of Morris and Therivel (1995). Some methods can be directly used for ecological risk evaluation.

1.5.2. Resource Management

The strategies of resource management were developed by managers to estimate the impact of exploiting natural or semi-natural resources. For example, the strategy indicated by Walters (1986) for managing the numbers of fish in a pisciculture:
- develop a model of the resource;
- estimate the parameters of the model;
- exploit the model from the data known about the resource;
- test the model in nature for some years;
- revise the model and its estimated parameters from the results obtained.

These models are applicable to accidental chemical pollution, assimilating the mortality caused by the pollutant to the sample taken by the manager. Other bibliographical references on the application of models to resources can be found in the work of Suter (1993a).

1.5.3. Evaluation of Danger

Danger evaluation is quite an old technique (Cairns et al., 1978). It is based on a sequential and iterative process of comparing two values that characterize environmental concentrations (measured or estimated) and the threshold of danger (measured or estimated). If environmental concentrations are lower than the threshold of danger, taking into account the respective intervals of confidence of these two values, one can take a positive decision to accept the risk. In the contrary case, one does not accept the risk (decision to reject) or one demands supplementary data to reduce the intervals of confidence, repeating the process until a decision can be taken. The differences between danger evaluation and risk evaluation have been specified by Suter (1990a), but this distinction is not always very clear in practice.

The rather general approach that consists of establishing predicted environmental concentration (PEC) and predicted no effect concentration (PNEC) uses a practice of danger evaluation, but not risk evaluation (Calow, 1993). However, it would be better to acknowledge that there is danger evaluation if the evaluation is based only on intrinsic properties of the molecule, and risk evaluation if one also considers exposure—which is the case here. The PEC/PNEC approach is very useful in obtaining prospective estimates (cf. section 2.2) on the regional scale but is less significant at the local and retrospective level, which is the case with several polluted sites (cf. Annexure 4).

1.5.4. Development of Rating Systems

Rating systems are designed to establish priorities. They have typically been used for the ranking of polluted sites and the definition of decontamination priorities, but also to classify chemical products according to their toxicity, for example, SIRIS (*Systeme d'Integration des Risques par Interaction de Scores*, or System of Integration of Risks by Interaction of Scores; Jouany et al., 1982). Suter and Barnthouse (in Suter, 1987) describe several rating systems, such as that used by the US Department of Defense (Smith and Barnthouse, 1987), that evaluate human and ecological risk from estimated dispersion of pollutants on the surface or in layers of the water table. In these systems, various constituents are described by attribution of intermediate scores (from 0 to n), then these intermediate notes are combined according to various modalities to obtain a final score.

The artificial nature of rating systems is evident. The attribution of scores presupposes clear-cut categories and the decisions are difficult. Simple rating systems are very useful, but there is a great temptation to complicate them, and the value of classifications obtained under these conditions may then be questioned. The Hazard Ranking System of the EPA (1982 and 1990) is such a rating system that is strongly criticized because of a certain number of biases that falsify the results (Doty and Travis, 1990). As an example of such bias, Suter (1993a) remarked that the effects are scored on the basis of the quantity of the most toxic product present on the site, but the total quantity of pollutants is scored on the basis of the combination of all the compounds present, so that sites with a large quantity of slightly toxic materials and a small quantity of very toxic products are attributed very high scores.

2. GENERAL CONCEPTS

2.1. Two Simple Examples

The following two examples are representative of situations that occur in practice for the risk evaluator, the first related to health risk and the second to ecological risk. We present them in order to better bring into relief the points common to the two operations.

Example 1

The drinking water of a city is contaminated by pollutant X, for example, lead. What is the health risk for inhabitants of the city? The risk evaluation will go through several stages:

- Measurement of human exposure (degree of contamination of the population): What is the concentration of the product in the water? What quantity of water will be consumed? By what category of the population (man, woman, child)? During what time period?
- Estimation of the potential danger: The data found in the literature give reference doses (RfD), that is, the quantities that may be absorbed during a lifetime without risk to health. These doses are calculated from dose-toxicity relationships established by experimentation on animals, generally after division by a security factor (or factor of application, cf. chapter 3, section 4.2), designed to take into account the differences between the actual situation and the existing data. The further these data are from the actual situation, the larger the security factor. The security factor is very large if there is only a single datum, for example, a single LD_{50}[11] (cf. chapter 3, section 3.1) for rats. Conversely, if long-term studies are available on the sub-lethal effects of the product, or even epidemiological studies analysing episodes of exposure of human groups in conditions similar to those in the case being studied, the security factor is smaller by several orders.
- The risk will finally be characterized by a simple method, in which the exposure is divided by the RfD (quotient method); if the ratio is less than 1, there is no risk, and if it is greater than 1, there is a risk.

Example 2

A site containing industrial waste, mostly cadmium, pollutes the water of a nearby water course lying below the site. What is the ecological risk for earthworms in the soil and the fish in the water? The approach is as follows:

- Measurement of the pollutants present on the site, in the soil and water.
- Research in the literature on cadmium toxicity to these species; generally laboratory ecotoxicity tests provide doses without effect (for example, NOEC or LOEC[12]). Otherwise, these data can be generated in the field under study, by laboratory bioassays, establishing

---

[11]LD_{50}: Lethal dose 50, the dose causing 50% mortality in a sample population.
[12]NOEC, no observed effect concentration; LOEC, lowest observed effect concentration; cf. chapter 3, section 3.1.

directly the dose-response relation of the pollutant for the species concerned.
- Characterization of risk by the quotient method.

These two examples are very simple and it is easy to see that the approach is the same in the two cases:

- Formulation of a problem of environmental toxicity, with two factors, a pollutant and one or several contaminated organisms.
- Definition of dose of exposure.
- Use of experimental or epidemiological data establishing a relationship between dose and toxic effect.
- Characterization of the risk: Use of the preceding relation to calculate the significance of the toxic effect (or its absence) as a function of the exposure dose.

2.2. The Ecotoxic Process, Scenario and Models

The appearance of toxic effects is the consequence of a series of successive actions that occur in a determined spatial frame, and that place two entities in contact with each other: the *pollutant* (dangerous object) and the *polluted* (the biological systems at risk, that is, the biotic or abiotic elements that will be disturbed by contact with the pollutant). These sequential stages are represented in Figure 1.1:

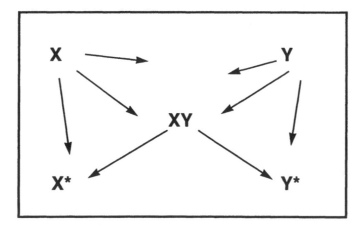

Fig. 1.1. General diagram of the ecotoxic process:
X, pollutant (substance); Y, polluted (biological system); →, forces that activate X and Y, ensuring their movement in the biological system (large framework) and their interaction (XY); (*), consequences of this interaction on the pollutant itself (degradation, metabolism; X*) and on the polluted (adverse effects; Y*).

• Emission of a pollutant from a source
• Occurrence and action of the pollutant on the environment
• Contact with elements at risk
• Appearance of adverse effects in the elements

The qualitative and quantitative description of this causal chain will contain the following points:

• Nature, localization, duration, intensity, and frequency of the emission by the source
• Spatio-temporal evolution of the pollution in the environment
• Spatio-temporal evolution of the elements at risk
• Nature, duration, intensity, and frequency of the interaction with these elements
• Nature, duration, intensity, and frequency of the adverse effects that result from the exposure.

The ecotoxic process occurs in principally two illustrative cases. In one the pollution is only potential, for example, a new pesticide, the effluents of a new factory, or the establishment of a new excavation site. The second situation corresponds to an existing pollutant, for example, the presence of an incinerator, a factory effluent, or a polluted site (or the supposed existence, for example, of a water course at a point lower than a discharge). The distinction between these two situations was neatly underlined by Suter (1993a) when he proposed two types of evaluation, a *prospective evaluation* and a *retrospective evaluation*.[13] We use these two terms[14] to designate the two illustrative cases that will be described, noting that the objectives of prospective evaluation are clear, concerning the prediction of potential effects of a pollution that does not yet exist. Retrospective evaluation has more ambiguous objectives, since it is concerned with the revelation of present effects, more or less visible, of an existing pollution, but also the long-term development, which is a predictive operation. This distinction is important. In the field of ecological evaluation of polluted sites, as defined by Warren-Hicks et al. (1989), it is specifically mentioned that the ecological effects evaluated are past and present effects, but not future ones. The Canadian approach is not so categorical in this regard: Gaudet et al. (1994) state that 'the future is very significant even in seeking to evaluate an existing contaminant'.

Whatever the defined objective, retrospective evaluation differs from purely prospective evaluation, because it incorporates an epidemiological component. The *causal chain* (the *chain of risk*, according to Merkhofer, 1987) must be established in a single direction in the case of a prospective

---

[13]Suter defined also a minor category, *retrodictive evaluation*, as a retrospective evaluation that resembles a prospective evaluation because very few of the elements are known; he cites the example of the dispersal of a relatively biodegradable pollutant in a marine environment.
[14]They correspond to the *prognosis* and *diagnosis* used by Verhoef and van Gestel (1995).

evaluation (from the source to the effects) and in two directions (from the source to the effects and from the effects to the source) in a retrospective evaluation. Some authors use the terms *bottom-up methods* and *top-down methods* (Gaudet et al., 1994).

In a risk evaluation, the objective is not to describe the situation exactly as it is, but to make a *model*, in the first sense of its definition,[15] that is, a simplified representation of events that occur or may occur, as a function of a scenario[16] decided during the initial phase of *risk* formulation. The scenario identifies the environmental problem to be solved and indicates and specifies the predicted causal chain:

- the spatial and temporal framework
- the factors present (X, the pollutant, and Y, the polluted)
- the forces that activate them ($\rightarrow$)
- their interaction
- the consequences of this interaction (*) on the pollutant itself (degradation, metabolism; X*) and on the polluted (biological effects; Y*).

The model is then *constructed*, that is, the logical elements (qualitative and quantitative) necessary to the realization of the scenario are put into place. They are then made *operative*, from qualitative or quantitative givens (*parameters*). The missing data are replaced with *extrapolations* or *hypotheses*. One of the principal extrapolations consists of passing from the final point of measurement to the final point of evaluation. For example, in the case of the illustrative figure cited in section 2.1, if one supposes that the fish present in the water course are salmon and that the toxicity is characterized by laboratory assays on trout, one will use certain hypotheses or develop particular methods of extrapolation (cf. chapter 3) to adapt the toxicity values of trout to salmon; this adds to the total uncertainty of the result.

To conduct a risk evaluation, different operational tools are needed.

2.3. The Operational Tools: Approaches and Methods

The use of a model is dependent on the operational tools (the methods, as described by Covello and Merkhofer, 1993) at one's disposal.

---

[15]The word *model* can have many meanings. It is a 'formalized structure used to take into account a combination of phenomena that have certain relationships to each other; that which is given to serve as reference' (Petit Larousse, 1992). It can be a *mathematical model*: 'mathematical representation of a physical, human, etc., phenomenon, realized in order to better study it' (Petit Larousse, 1992). Lastly, we later use the term *physical model* to designate a simplified material representation of a complex biological phenomenon.

[16]Scenario: 'predicted occurrence, programme of an action' (Petit Larousse, 1992). The term *scenario* is not used by Covello and Merkhofer (1993); for Suter (1993a), the scenario is limited to the exposure phase. Here, it will be employed in its widest sense, encompassing the entire operation.

First of all, a risk evaluation is initiated with a minimum of preliminary information indicating the possibility of a risk and the necessity of making this evaluation: the proven or potential presence of pollutants, visible adverse effects of these pollutants on the environment or living organisms, etc. Risk evaluation is then done from *existing data* (for example, the results of ecotoxicity assays published in the literature, in general the LD_{50} or NOEC), from *data generated for the purpose* (for example, the results of bioassays or eco-epidemiological data[17]), and methodological tools, such as mathematical models. The different methods that can be used correspond to three types of approaches, testing, monitoring, and modeling (Chapman, 1989, 1991).

- *Experimental models*: these are the *conventional assays*[18] of the occurrence, behaviour, and effects of pollutants (1) at different levels of organization, that is, laboratory assays (*monospecific tests*) and different integrated assays (from multispecific assays up to mesocosms) and (2) for different types of pollutants, from the pure substance to the polluted medium (*bioassays*).
- *In situ indicators*, relative to the environment and to the living elements that populate it or living elements introduced on the site (measurement of pollutants and eco-epidemiological data).
- *Mathematical models*.

For Covello and Merkhofer (1993), this categorization is not absolutely rigid; for example, some mathematical models may be used to express results of conventional assays. Moreover, the distinction between a laboratory assay and an *in situ* indicator is fundamentally arbitrary. An ecotoxicity assay is a microcosm, in the first sense of the term, that is, a 'world in miniature' that attempts as far as possible to represent the complexity of nature, while the data collected on the site can be considered the result of a single experiment on a grand scale. The observation itself presupposes an experimental part (definition of a collected protocol of living elements) and the introduction of living elements on a polluted site is also a form of experimentation. The bioassays have been linked to ecotoxicity assays, because they have been conducted according to the same standard protocols, but they can be considered a particular form of *in situ* indicators, since they use a polluted medium rather than a pure substance.

The value of these different approaches is variable. The choice is dictated by the data available or the possibility of generating additional data, and the type of situation to be evaluated. The environmental effects of a new pesticide are predicted essentially on the basis of standard assays and ex-

---

[17]Here we distinguish epidemiological data, relating to human populations, from eco-epidemiological data, relating to animal and plant populations.
[18]The regulatory terminology uses *assay* rather than *test*.

periences of limited scope on the terrain, while the evaluation of a site or soil polluted by pesticides should take into account *in situ* indicators, such as the concentration of the substance in the mediums and, finally, the modifications of plant and animal populations.

2.3.1. Experimental Models

Experimental models correspond to the 'physical models' of Suter (1993a). They are physical or biological systems simulating under controlled and simplified conditions the progress of the whole or part of an ecotoxic process. The determination of the coefficient of octanol–water share (K_{ow}) of an organic product is one such model. A certain quantity of the substance is introduced in a well-delimited environment (a phial containing two non-miscible liquids, octanol and water); after shaking, the concentration of the product is measured at each phase and the ratio is calculated. This experiment is a very simplified model of the process of transfer of molecules between an aqueous environment and a lipophilic environment; the model is only *partial*, because it does not indicate the potential effects on living elements.

In another example, in order to determine the acute toxicity of a pesticide on trout, an ecotoxicity assay is done: a fixed number of fish are introduced in a limited environment (an aquarium filled with a known volume of water), and some milligrams of pesticide are added. After 24 or 48 hours, the dead fish are counted. The events that occur during the assay are basically the same as those that occur in nature, but the characteristic phases of the process occur in a short time and in a simplified fashion. For example, the phase of emission and the phase of exposure are brought to their simplest expression: the introduction of a certain quantity of the product in the aquarium at time t stands for immersion in a polluted body of water or consumption of a certain quantity of contaminated feed. Ecotoxicity assays are thus sufficiently good physical models, but they must be augmented by other elements that enable us to evaluate more precisely what happens in nature, for example, interspecific extrapolation models (cf. chapter 3).

The final expression of an ecotoxicity assay can take different forms: direct expression of the desired result (mortality of a certain number of fish at each concentration tested) or a mathematical model linking the numeric variables. The result may be communicated in various forms, such as a graphic representation of the percentages of mortality as a function of the dose, a curve adjusted on these data (a line representing the log(dose)–integer relation, for example), or even a single value (LD_{50} or NOEC).

Monospecific laboratory assays with pure products administered to some laboratory species are the most widespread form of experimental assays, but it is possible to create larger and more complex models, ranging from multispecific assays to mesocosms (integrated assays). The results obtained with these models are more difficult to use for the evaluation of ecological risk.

The authors of I2C2 (1994), with good reason, distinguish two categories of tests: *conventional tests*, standard or not (routine assays), and *parametric tests*. Conventional tests serve as the basis for risk evaluation, while parametric tests serve to extrapolate the values of standard tests to other situations, for example, to adjust the results of a conventional test, conducted under a determined temperature, to the range of temperatures found in natural conditions (cf. chapter 3, sections 2 and 4.3.9).

The advantages and disadvantages of conventional laboratory tests are well known (cf. chapter 3); their chief advantage is that they are reproducible. They are generally cheaper and quick, but they have little 'ecological realism'. Besides, they are not indispensable, as they contain much that is only a model of the elements that constitute the scenario.

Bioassays are experimental devices designed to measure the effects of the mediums from a site under laboratory conditions. They are particularly well adapted to the evaluation of polluted sites. Most bioassays are done in conditions identical to those of conventional ecotoxicity assays.

2.3.2. In Situ Indicators

In situ indicators are:
- The measurements taken on the site to determine the concentration of pollutants.
- Eco-epidemiological observations designed to bring toxic effects to light: the use of test organisms and the observation of toxic effects on cultivated plants or caged animals on the site also belong to this latter approach.

These two types of indicators are to be found in natural ecosystems or in manipulated ecosystems (cf. chapter 3). The nature, advantages, and disadvantages of the different indicators and their use in risk evaluation protocols will be discussed later.

2.3.3. Mathematical Models

Mathematical models are divided into two main categories:
- statistical models
- mechanistic models (deterministic or stochastic)

The statistical models have three principal applications in risk evaluation:
- to test hypotheses
- to describe events and phenomena
- to extrapolate

Tests of hypotheses have been used, for example, in the evaluation of contaminated sites, to compare polluted sites to reference sites. Their use has been criticized because of the problem of interpretation: tests of hypotheses are generally based on the rejection of null hypothesis. The null hypothesis signifies that there is no significant difference between the two situations (for example, the polluted site and the reference site) and to reject this hypothesis is to say that there is a difference. Two types of errors are conventionally associated with these tests. The first type of error is the rejection of the null hypothesis even when it is true (we see a difference even when there is none) and the second type of error is the acceptance of the null hypothesis even when it is false (we do not see a difference even when there is one); α is the probability of making an error of the first type and β is the probability of making an error of the second type. The validity of the test is defined as $(1 - \beta)$. In most cases in which the test of hypothesis is used, the probability of making the first type of error is sought to be reduced as far as possible, with the reasoning that to conclude (wrongly) the absence of difference (that is, accept the null hypothesis) is preferable to the reverse. It is a prudent approach when we don't want to conclude too quickly (and erroneously) about the efficacy of an amendment, but in the case of a toxin, we risk concluding (wrongly) that there is no effect. In the case of comparison of two polluted sites, it is better to be mistaken in concluding a difference, that is, that a site is polluted even when it is not, than in concluding that it is not polluted even when it is. A more detailed discussion of the possible consequences of deficient validity of tests and the consequences of that for risk evaluation of pollutants can be found in Forbes and Forbes (1994).

Statistical models also contribute to the description and interpretation of test results, for example, the classic log(dose)–integer that links the concentration of the toxin to mortality. A more detailed presentation of various statistical models can be found in Covello and Merkhofer (1993).

Finally, *statistical models* (regression models) are the source of algorithms that serve to extrapolate, for example, from the tested species to the species present in the natural environment, or to doses that are outside the range tested, or even to different products (cf. chapter 3).

Stochastic models are based on the uncertain character of events. These models, based on years of regular measurements, are well adapted to meteorological predictions, for example, or predictions of automobile accidents, but they necessitate a very large quantity of data in order to be useful.

Deterministic models correspond to those generally spoken of as models, that is, a mathematical formalization of relations between the different elements of the system, based on the description of physical, chemical, and biological phenomena. The two general types of models are models of *occurrence* and of *behaviour*, which simulate the occurrence and transfer of products in the environment (cf. chapter 2) and the models of effects at different levels, at the organism level (toxicodynamic models, for example) or at the population level (cf. chapter 3).

The *validity* of models developed for the evaluation of ecological risk is very often disputed. As Suter (1993a) has shown, part of the difficulty arises from an insufficiently precise definition of what is understood by validation, which can be stated as follows:

- the model corresponds exactly to reality;
- the model has made satisfactory predictions.

The first is much too absolute. A model must have been verified in some specific cases, but by definition, it is designed to evaluate a situation that has no precedent. According to Covello and Merkhofer (1993), a model must always be 'false'. The only means of ensuring its validity is to apply a great number of different experimental situations and to verify that the results conform to this model. This is possible for the modelling of small, relatively simple systems corresponding to a situation of small amplitude (for example, the transfer of herbicides of the same chemical family in corn leaves), but not feasible when the models increase in size and complexity because of the time and space that would be required.

Other criteria of validation have been suggested, such as publication in specialized reviews, with a reading committee. Consideration in regulatory norms is often cited as a proof of validity of models. In fact, it means simply that the models are the object of a general consensus (or have been imposed), but that does not mean that they are the best adapted to the situation, or the most scientifically founded.

Theoretical models have been proposed to guide the selection of models, but they are not often used (Suter, 1993a; Covello and Merkhofer, 1993).

2.4. Choice of Model

The choice of model obviously depends on the chosen scenario, but in fact, the possibility of realization of a scenario is very dependent on the available models. The model is generally constructed from existing submodels or those generated during the course of evaluation, constituting the different links in the causal chain. These givens are rarely homogeneous and the art of the evaluator lies in adjusting the different links between them. This construction of the model from very disparate elements is characteristic of risk evaluation; the model is a composite, according to the definition of Covello and Merkhofer (1993).

The scenarios, like the corresponding models, are simple or complex, partial or total.[19] The general outline presented in Fig. 1.1 is an example of a scenario that is simple and total, but not usable, because it is too abstract, the elements that make it up not being defined precisely and not capable of being quantified.

Many models are only *partial*, representing only the exposure phase or one part of it. For example, there are models that describe only the occurrence of a product or its biotransformation in an environment; others characterize only the means of exposure. Models gain *overall* in considering more various situations, for example, in incorporating a larger number of stages. A model linking the environmental concentration to the *internal dose*[20] will be more total than a model linking the environmental concentration to the external dose. In order to construct the definitive model, it is customary to combine several partial models, for example, a partial model describing the occurrence of a product will be associated with a model describing the movements of populations at risk.

The *complexity* increases if the content of the different steps is more detailed. For example, we augment the description of the spatio-temporal evolution of a product by that of its metabolites, or even, instead of reducing the population to a standard individual of the species considered, we take into account a population composed of individuals of different sensitivities (young people and adults, for example).

The choice between a simple and complex model, and between a total and a partial one, depends on the objects of the study indicated in the scenario and the significance of the necessary data; simple and total models will suffice for a rapid evaluation of risk related to a chemical product. In example 2 in section 2.1, we can use a simple and total model. The source of emission is regular, homogeneously polluting a defined geographic zone (in the case of a polluted site, we acknowledge that the concentration in the soil is the same at all points at a determined depth). The exposure will be that of an 'average' individual of the species considered (an earthworm, for example), present in the polluted zone during its entire lifetime.

The detailed scenarios necessitate elaborate, complex models and a considerable number of data that it is not always possible to obtain, which can lead to several decisions:
- generate the models and missing data by specific experimentation or by extrapolations;

---

[19]The term *global* would be preferable, but it is already used to designate scenarios of continental or global geographic scale.

[20]We use the general terms *external dose* and *internal dose* to designate, respectively, the quantities or concentrations present in the environment (in contact with the organism) and present in the organism.

- define a more simple scenario, possibly by redefining the *final points* (cf. section 3.2.2), and construct a model less demanding in terms of data or making better use of the available data.

A complex model is not always indispensable. The height of complexity is an analytical model, detailing all the possible situations, but such a model—even if it were possible—is of little use. The essential problem is not to study the entire ecotoxic process to its smallest detail. The integral understanding of the molecular mechanisms is not indispensable to linking the doses and the toxic effects. Ecotoxicity assays shorten the intermediate steps and establish direct relations between external doses (environmental concentrations) and toxic effects, without the necessity of knowing the internal doses. Bioassay results and the existence of eco-epidemiological data also ensure a direct link between the environmental concentrations— more rarely the internal doses—and toxic effects. On the other hand, the use of biomarkers demands that there be an established relation between the dose of the pollutant and the response of the biomarker, on the one hand, and between the response of the biomarker and the toxic effect, on the other hand.

Risk evaluation must be *flexible* and *iterative*. Some new final points will be incorporated during the course of the evaluation if it is felt to be necessary. It is also possible to proceed in stages: first a summary characterization is done, and if the results show a certain degree of risk, a more detailed evaluation would be needed.

The scenarios are simplified representations and compromises, which is why there are several possible scenarios that attempt to describe the same situation, and, as a corollary, *different results*, independent of the uncertainty associated with the parameters of the model (Dobson, 1993; Nillson et al., 1993). Suter compares risk evaluation to what happens in a court: there is a presumed culprit, the pollutant, and a presumed victim, the polluted. The court (risk evaluator) will use all possible means to attempt to reconstruct the sequence of events (the scenario) as exactly as possible by the presentation of material proofs, such as confessions, expert techniques, etc. (the different approaches and methods of risk evaluation). This comparison is interesting, because it brings to light the decisive importance of human judgement. For any case, it is well known that the verdict depends on the judge and jury. Risk evaluation also, with the same data and the same material elements, can progress according to different scenarios depending on the options selected by the evaluator and does not always end in the same conclusions. The differences can be very large, whence the acknowledged necessity of a large professional experience and a significant weight to human judgement. But contrary to a process in which there are no absolute proofs of the variability of the verdict (one cannot commit the same crime twice), it is possible (at least theoretically) to develop various scenarios and verify the one best adapted to the actual development of the situation.

This example shows also the significance of *expert judgement*,[21] representing the state of the evaluator's understanding. Suter (1993a) remarks that the results obtained by the judgement of experts are not necessarily worse than those from a model based on more scientific data (a mathematical model, for example), but there are two disadvantages:

- the procedure is less transparent to others;
- experts have the tendency to have an exaggerated confidence in the value of their evaluations, which biases the final result.

The credibility of a 'scientifically based' model could be better, but that supposes:

- the belief in a certain truth to science;
- that the model rests on true and verifiable scientific bases (in the sense of Forbes and Forbes, 1994).

Differing results are obtained depending on the type of model chosen, but also depending on the scenario envisioned. For example, to calculate the exposure of a population of a city to a pollutant present in fish, the scenario may consider the exposure of an individual consuming n grams of fish a week. These n grams are estimated by statistical studies of the average consumption of individuals. This average approach is not suitable to all cases, particularly when it is necessary to evaluate the risk to sensitive groups, for example, sub-populations that, for various reasons, consume much larger quantities of fish. This problem is resolved by explicitly incorporating sensitive groups in the exposure scenario or by defining the maximal rather than average values, in estimating the consumption of fish by the population. In another example, a phytosanitary treatment is done in variable doses according to the crop, and, rather than proposing different scenarios, the evaluation is based on the highest doses of the application to evaluate the overall risk. This strategy, called 'worst case', or even 'extreme case', is systematically used, but one must not forget that the risk evaluation must remain *reasonable*.

3. GENERAL CONCEPTUAL MODELS

The formalization of conceptual models is hardly fifty years old. The first to be developed were models to evaluate health risk, the prototype of which is the model established by the US National Academy of Sciences.

---

[21]The reader interested in the function of expertise and the role of the expert in decision-making can consult the excellent introduction of Roqueplo (1996).

It is described in this section, along with a more recent model proposed by Covello and Merkhofer (1993).

There is no single, unanimously adopted model for ecological risk evaluation. The most important contributions, the model proposed by the EPA (Norton et al., 1992) and those proposed by Suter and his team in 1993, are described here. The same phases are found among these different models, because of the essential role of the concepts developed by Suter, which strongly influenced the EPA model. All these models are very general, characterized by a high degree of abstraction and formalism that makes them difficult to understand immediately. Two other significant models for evaluation of ecosystems, those of Barnthouse (1992) and Lipton et al. (1993), are briefly presented in this section. The principles of risk evaluation by the PEC/PNEC type of approach, as they are spelled out in the regulations of the European Community, are described. There are not many conceptual models specifically adapted to contaminated sites, and two examples are explained here, the EPA model (Warren-Hicks et al., 1989) and the Canadian model (Gaudet et al., 1994).

3.1. The Model of the National Academy of Sciences

The basis of risk evaluation was established in 1983 by the US National Academy of Sciences, which defined risk evaluation as the scientific activity that consists of evaluating the toxic properties of a chemical product and the conditions of human exposure to this product, with a view to observing the reality of a human exposure and characterizing the nature of effects that can result from it (National Research Council, 1983).

The process consists of three preliminary stages (Fig. 1.2):

*   *identify the danger* (determine whether a chemical product has effects, understood to mean harmful, on health);
*   *evaluate the dose-response relation* (determine the relation between the level of exposure and the probability that the adverse effect in question will appear);
*   *evaluate the exposure* (determine the level of human exposure, in different conditions).

Corresponding to each of these stages is a phase of research that collects existing data from earlier studies or data specifically generated for the study. These are results of toxicity tests on animals (LD_{50}, NOEL[22]), measurements of environmental concentrations of pollutants, and epidemiological data, if any, on populations exposed to pollutants. The results of the three

---

[22]NOEL = No Observed Effect Level, syn. NOEC.

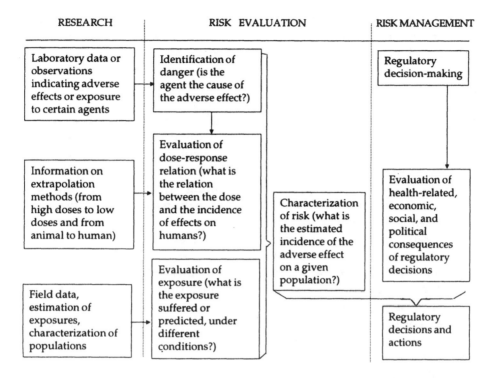

RESEARCH | RISK EVALUATION | RISK MANAGEMENT

Fig. 1.2. General outline of a health risk evaluation: the model of the National Academy of Sciences (National Research Council, 1983).

preceding operations are combined to *characterize the risk*, that is, describe the nature and level of risk for the individual in a given human population.

This primary outline rapidly evolved until, in 1986, the Task Force on Health Risk Assessment (DHHS, 1986) recognized four stages, a little different in their designations and content:

- *identification of danger*: qualitative indication that there exists a substance or situation that has harmful effects on human health;
- *characterization of danger*: evaluation of the nature of adverse effects and their expression as a function of exposure (dose);
- *characterization of exposure*: qualitative and quantitative evaluation of probable degree of human exposure;
- integration of the preceding stages in a scientifically based determination of risk level as a basis for regulatory decision-making.

All models of health risk evaluation tend to fix on a general consensus recognizing these different stages; the stages are not explained here, except for the Covello and Merkhofer model (1993) of health risk evaluation, which is presented in comparison with the NAS model (Fig. 1.3). It is more recent than the latter and differs from it in numerous points:

- The NAS model considers identification of danger to be the first stage in risk evaluation, whereas Covello and Merkhofer consider it a preliminary stage, to be realized before risk evaluation itself. For these authors, it is a fundamental stage from which to determine causes and effects and characterize causal relations: almost half of the evaluation is already done in this stage.
- The NAS model does not include the stage of identification of sources, whereas the Covello and Merkhofer model does. This definition of a separate stage is justified by the particular importance accorded by these authors to *industrial accidents*. They remark, rightly, that a prediction of failures of manufacturing systems (factories) or transport systems (e.g., long distance canals, road transportation) is essential to obtain a complete evaluation of the risks of a product. It can also prove useful in a retrospective evaluation attempting to identify the sources of pollutions and thus ensure that the rehabilitation will be complete, without risk of further pollution.
- The next stage, evaluation of dose-response relation (NAS) or evaluation of consequences (Covello and Merkhofer), covers the same operations, except that the Covello and Merkhofer model incorporates the environmental dimension, while the NAS model is strictly designed to evaluate health risk.
- The evaluation of exposure, which comes after the evaluation of dose-response relation in the NAS model, is placed before the evaluation of consequences in the Covello and Merkhofer model, which conforms more to the notion of a causal chain.
- The last stage is named 'estimation' by Covello and Merkhofer, rather than 'characterization', to underscore the fact that 'risk is a characteristic of the world in which we live and that the objective of risk evaluation is to communicate estimations of risk in comprehensible terms, and not to produce abstract data from a model.'

3.2. The Ecological Model of the EPA

3.2.1 General Characteristics

The model proposed by the EPA (Norton et al., 1992; Fig. 1.4) has a very general scope, in that risk evaluation is conceived at the outset as 'an operation that evaluates the likelihood that adverse ecological effects are produced as a result of exposure to stresses.' The term *stress* here comprises a wide range of physical, chemical, or biological effects resulting from human activities and capable of disturbing the components of ecosystems, that is, individuals, populations, and communities. The effects considered may range from the death of an individual to the destruction or loss of

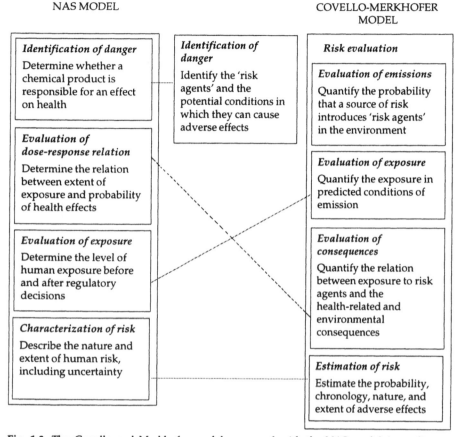

Fig. 1.3. The Covello and Merkhofer model compared with the NAS model (according to Covello and Merkhofer, 1993; with the permission of Plenum Press, 1997).

function of an entire ecosystem. The EPA model continues to be based on models of health risk evaluation, but emphasizes three points:

- Effects must be considered at higher levels than those of an individual of a single species and the possibility of effects on populations, communities, and ecosystems must be included.
- It is not possible to rely on a single set of ecological values, the protection of which is agreed on beforehand.
- Models must be developed that are valid not solely for chemical stresses.

The example used by the authors to illustrate this model is the supposed impact of supplementary feeds released into an estuary, which underscores the importance given to stresses other than chemical stresses. This model is designed for a prospective evaluation as well as a retrospective one.

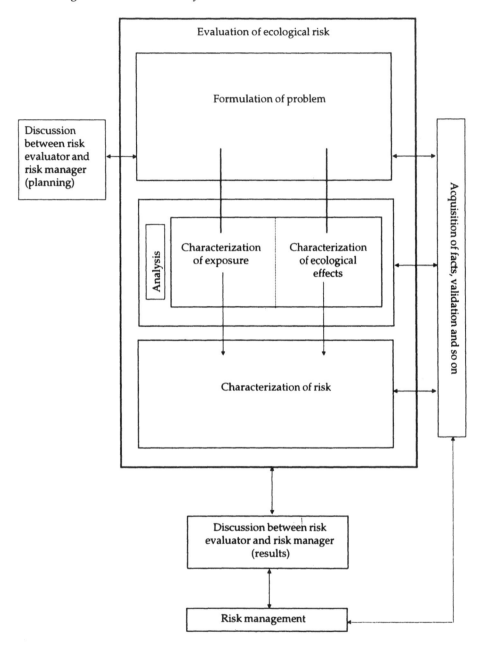

Fig. 1.4. General outline of ecological risk evaluation: the EPA model (Norton et al., 1992; with the permission of SETAC, 1997)

Risk evaluation is *iterative*. For example, supplementary data may be needed at any time: the process comes to a halt, and the necessary data are obtained and incorporated. The interaction between the risk manager and the risk evaluator is constant. At the outset, the evaluator must ensure that all the 'ecologically relevant' questions are posed by the manager, while the manager must agree with the evaluator on the nature of the final points (see section 3.2.2). Risk evaluation is not considered a directive line but a conceptual field defining a relatively homogeneous methodology and terminology, designed to be consequently adapted to very diverse situations of prospective and retrospective evaluation. This is to say that ecological risk evaluation is—presently—a conceptual field more flexible than directive lines. On this subject, Forbes and Forbes (1994) make the interesting remark that directive lines—like those of the OCDE's ecotoxicity assays[23]—designed at first to be relatively flexible propositions, are quite rapidly transformed into rigid protocols, since there is a great tendency on the part of regulatory authorities to rely on standards and norms. We must remember that risk evaluation procedures presented today as rather informal outlines may tomorrow become demanding regulations. Risk evaluation of pesticides developed in this manner and we must be very careful that, some day, the same does not happen with other operations, such as ecological risk evaluation of polluted sites.

The model proposed by the EPA puts into play a series of operations summarized in Figure 1.4. Each is divided into sub-stages.

3.2.2. Formulation of the Problem

The problem is presented by the *risk manager*, who can be a regulatory authority, a unit responsible for polluting with its waste (smokestacks, effluents), or the victim of pollution (for example, a pisciculturist). During this stage of planning and delimitation, the ecological risk evaluation unit, represented by the *risk evaluator*, has the objective of preparing a conceptual model (the scenario). This model identifies:
- the stresses, the ecosystems to be protected, the elements at risk (definition of final points of evaluation);
- the spatial and temporal scales of phenomena, the approaches used (e.g., toxicity tests, bioassays);
- the necessary factual data.

This model helps in the selection of the *final points of measurement* that are the substitutes for the *final points of evaluation*. The distinction between the two is clearly defined: 'the final points of evaluation are explicit expressions of environmental values to be protected [elements at risk]', whereas 'the final points of measurement are responses to stresses,

---

[23]OCDE = Organisation de Coopération et de Développement Économique.

measurable and relative ... to final points of evaluation.' This distinction is necessary because the final points of evaluation are difficult to measure directly. For example, the final point of evaluation may be the survival of a population of pike: the evaluation is done by the intermediary of a final point of measurement,[24] generally the result of an ecotoxicity assay on a laboratory species (for example, an LC_{50}–96 h[25] test for trout). Very often, the final points of measurement at a lower level of organization will serve, for want of something better, to characterize a final point of evaluation at a higher level of organization. The EPA recognizes that a certain level of expert judgement is necessary for a rational selection of these different final points, an essential condition for a good evaluation. This model also includes a *scenario of exposure*, that is, a qualitative description of the relations between the stress and the ecological components. In the example presented by Norton et al. (1992), potential impacts resulting from the flow of excessive nutrients are envisaged. In the case of fish, benthic communities may be disturbed, rates of dissolved oxygen will diminish, and ultimately the survival of fish populations will be compromised. If one wishes to evaluate impact on aquatic vegetation, in a preliminary formulation one concentrates on maintenance of the abundance and spatial distribution of several plant species that will constitute the final point of evaluation. In all the cases illustrated, it is all right to modify the objective at any point in the operation and to incorporate other final points of evaluation. The final points of measurement will be the growth and reproduction of these different aquatic organisms or acceptable substitutes (neighbouring species).

This initial phase of problem formulation serves as the basis for the next phase, *analysis*.

3.2.3. Analysis

Analysis comprises two parallel operations, *characterization of exposure* and *characterization of ecological effects*. The two are in constant interaction because the evaluator is confronted with a set of heterogeneous data, some missing and others not very reliable. A certain degree of expertise is necessary.

Characterization of exposure

Characterization of exposure consists of determining the possibilities of spatio-temporal contact between the stress and the receptor, on the basis of chemical analysis or modelling of environmental concentrations. Also,

---

[24]The concepts of final points of measurement and evaluation are explained in sections 3.3.1 and 3.4.1.
[25]LC_{50}–96 h: lethal concentration for 50% of the individuals, measured after an assay lasting 96 h.

the exposed populations are defined, as well as their points of contact with the pollutant.

Characterization of ecological effects

Characterization of ecological effects is based on toxicity and ecotoxicity assays conducted on different animal or plant species, or even on eco-epidemiological data from domestic or wild species exposed to environmental pollution (*sentinel organisms*). These values and data are found on the basis of existing data. They may also be extrapolated (with the use of mathematical models) or even generated in the field under study by specific experiments.

3.2.4. Risk Characterization

The results of the two preceding phases are combined according to various methods:
- comparison of unique values for exposure and effects;
- comparison of distributions of exposure and effects;
- use of simulation models.

The ecological risk will finally be expressed in different ways, in qualitative (absence or otherwise of risk), semi-quantitative (e.g., low, medium, or high risk) or probabilistic terms (as a percentage). A significant part of risk characterization will be to discuss the origin and extent of various uncertainties associated with the result.

Finally, the set of data is communicated to the risk manager, who takes a decision, generally after a risk–benefit analysis.

3.3. The Ecological Models of Suter

In contrast to the EPA, Suter (1993a) proposed two distinct models for prospective and retrospective evaluation.

3.3.1. The Prospective Model

The first outline proposed by Suter (1993a) pertains to prospective risk evaluation, that is, situations in which the effects of a new chemical product (e.g., a pesticide) or the biological effects of the effluents of a new factory must be predicted. The different phases set out in this model are represented in Figure 1.5.

Identification of danger

Identification of danger comprises three parallel operations that are closely interactive: *selection of final points, description of the environment,* and *modalities of emission* (definition of sources).

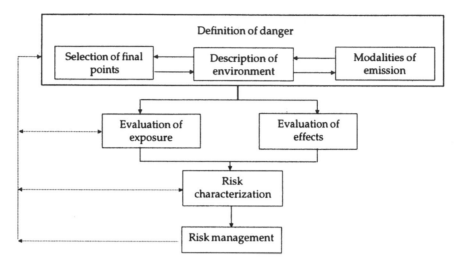

Fig. 1.5. General outline of prospective evaluation of ecological risk: the Suter model (Suter, 1993a; with the permission of Lewis Publishers, CRC Press, 1997).

Selection of final points is motivated by the need to answer two questions:
- What are the elements at risk?
- What are the predictable effects on those elements?

The first question corresponds to the stage of danger identification described by the National Academy of Sciences in 1983. Various approaches can be used in answering it:
- If the action is complex, one can create a matrix of agents (e.g., the effluents) and affected receptors (e.g., fish, birds, microflora); each possible interaction is indicated and if possible represented by a score based on exposure level and supposed sensitivity of the target.
- A rapid identification of the mediums or areas that will be the most polluted and the receptors at risk.
- Logical models, for example 'trees' of causal relations.
- The study of historical data on existing situations similar to the one studied (products of similar structure, for example).

Suter (1993a) emphasizes the distinction to be made between final point of measurement and final point of evaluation and the frequent confusion that occurs between the two (we have seen that this distinction is repeated in the EPA outline). Moreover, the operational definition of a final point comprises two elements—the *subject* (e.g., the species, the population at risk, or any other value) and a *characteristic of the subject* (e.g., the rate of reduction in population size)—which will be translated into numerical values. The correct approach is to define the final points of measurement according to final points of evaluation. For example, if the final point of evaluation is the possible mortality of salmon in a fishery following the discharge of polluted effluents, the final point of measurement can be LC$_{50}$–96 h for trout.

Description of the environment involves in the first place a representation of the environment under consideration, more or less concrete depending on the circumstances. The geographic limits of the evaluation may be defined at the outset, for example, the pollutant content in an effluent must not pass certain values at a certain depth in the soil. Other approaches are presently used, even if they are not scientifically justified, for example, conducting the evaluation within administrative limits (commune, department, or region).

The *modalities of emission* are estimations of spatio-temporal characteristics of the introduction of the pollutant in the system, from a source or a set of sources. The precision of these estimates is highly variable. The volume of effluents or incinerator fumes may be estimated at the outset with some precision. However, one will have a less precise idea of the content of a particular pollutant in incinerator fumes and an even less precise idea of the dispersal of mutagenic products. The significance of the source may be estimated by indirect measures, such as statistics on predictions of pesticide sales, or the quantities of products discharged (with all the imprecision that comes with such estimations).

Evaluation of exposure

Evaluation of exposure consists of several operations, in the first place to transform the estimations of emission values into estimations of values in contact with the final points. This involves the construction of a model of distribution, degradation, and transport in the medium based on chemical characteristics of the product (e.g., aqueous and lipophilic solubility, ionization, vapour pressure) and on the characteristics of the medium (e.g., speed of current in the case of discharge into a river). According to needs, these models may be very simple and firmly based on hypotheses often difficult to verify (for example, near an effluent, we presume that the concentration in water is homogeneous and one-tenth that of the effluent), or they may be more elaborate, as in a model based on hydrographic characteristics.

The second stage in evaluation of exposure is the evaluation of contact between stress and receptor. If we take again the example of an effluent and effects on a fishery, the animal populations at risk are easily identified. The situation is more complicated in estimating harmful effects on fauna and flora naturally present in a water course, because that requires precise identification of species present and a profound knowledge of the dynamics of their populations.

A particular problem requiring special treatment is the existence of an underlying pollution that adds its effects to those of the identified source (e.g., a geochemical pollutant or pollution of the same nature upstream).

We must be careful not to attribute the pollution of a water course to the existence of a polluted site when it is caused by contaminations upstream.

Evaluation of effects

Evaluation of effects aims to determine the relation between exposure and adverse effects on the receptor. Essentially, it is based on the results of ecotoxicity assays. We will see later in the retrospective model how to incorporate eco-epidemiological data in the evaluation. In principle, the approach is as follows:

- Completion of toxicity assays in different conditions of exposure (intensity, duration) to evaluate the seriousness and frequency of the toxic effect (e.g., mortality after 96 h exposure to given concentrations of a chemical product).
- Creation of a representative model of the results (e.g., the log(dose)–integer function) and expression in numerical form (e.g., LC_{50} values).
- Creation of models of effects that establish the relation between these values (final points of measure) and the final points of evaluation identified at the beginning of the operation.

Risk characterization is the final product of risk evaluation. It involves the integration of data about predictable exposure and values of dose–effect relations to obtain an estimate of levels of predictable effects from exposure. This can be achieved—at least theoretically—by a representation in a space of four dimensions (Suter, 1993a), which would be:

- the concentration of the pollutant;
- the duration of exposure;
- the proportion of the population affected;
- the gravity of the effect.

3.3.2. The Retrospective Model

The retrospective model proposed by Suter (1993a) is quite different from the prospective model (Fig. 1.6).

Definition of danger

In the *definition of motivations*, Suter distinguishes three different situations that are the origin of three types of evaluation:

- *Source-driven assessments*, pertaining to pollutions that are past but identified, such as oil slicks and polluted sites.
- *Effect-driven assessments*, when the effects are observed in the field, for example, bird mortalities or acidification of a lake.
- *Exposure-driven assessments*, needed when highly contaminated sources of exposure (e.g., food) have been discovered. According to Suter, this case is relatively rare but will become more numerous when programmes of biological surveillance are intensified.

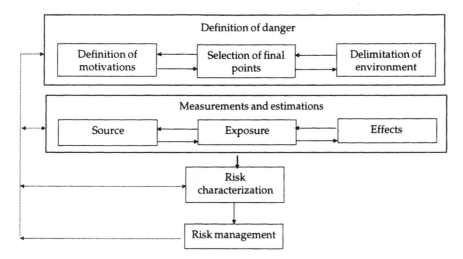

Fig. 1.6. General outline of retrospective evaluation of an ecological risk: Suter model (Suter, 1993a; with the permission of Lewis Publishers, CRC Press, 1997).

Suter's classification sometimes appears to be artificial: acidification of a lake may be considered as much a source-driven assessment (the 'acid' lake is a source of pollution, the origin of toxic effects) as an effect-driven assessment (the acidification is a consequence of acid rain or effluents). Suter remarks, rightly, that the *concept of source* is much more ambiguous in a retrospective evaluation than in a prospective evaluation. He defines the source as the problem of pollution posed by the manager, which is the object of the evaluation, and for which the evaluator attempts to link exposure and effect. For example, acidification of a lake may be linked to various causes, such as acid rain or acid effluents, which are potential sources. Similarly, toxic effects observed near a discharge or a polluted site may be due to this discharge or site, but not necessarily.

Delimitation of the environment is particularly critical, since the retrospective evaluation concerns a real situation. It can be based on the area of the zone in which the pollutant has been detected (source-driven assessment) or the area of distribution of the species affected or contaminated (effect- and exposure-driven assessments).

Selection of final points is based on the same criteria as for a predictive evaluation.

Measurements and estimations

When the source is identified, exposure is evaluated by conventional models of the occurrence and action of pollutants. If the source is not identified, there are methods for chemical characterization of the pollution at the point of exposure that enable us to trace it back to the potential sources (Gordon, 1988). The exposure may be directly measured from concentrations in the living elements present on the site or estimated from environmental concentrations of the pollutant and the response of biomarkers.

The effects are the variations in final points of evaluation attributable to pollutants. They are evaluated (in a predictive approach, using ecotoxicity assays), measured directly in the field or by bioassays, and eventually extrapolated, if the final point of measurement does not correspond to the final point of evaluation.

In fact, the different methods proposed by Suter in the *measurements and estimations* stage correspond—though that is not explicitly mentioned—to the *observation, experimentation, and modelling* described previously (section 2.3).

3.4. Other Ecological Models

Other models that have been proposed follow quite different outlines and have been developed in response to specific needs, such as risk evaluation for pesticides (Greig-Smith, 1992). Some are partial outlines, which do not account for all parts of the process (models of exposure, notably). Many models are conceptual propositions with a low operational capacity or very slightly different from the preceding models (for example, Lowrance and Vellidis, 1995). Two examples are described below, the models of Barnthouse (1992) and Lipton et al. (1993), and principles of risk evaluation such as those defined by the European Union are discussed.

The absence of a unanimously adopted terminology to designate the different stages of risk evaluation is a frequent source of confusion. Similar terms such as *analysis* or *evaluation* are used with quite different meanings. A streamlining of terms is needed to avoid redefining terms each time.

3.4.1. *The Barnthouse Model*

Like the EPA model, from which it differs essentially in terminology, the Barnthouse model has four principal stages (Barnthouse, 1992). The first stage is *identification of danger*, which determines whether a danger exists, for whom, and what supplementary information is necessary in order to evaluate it. The aim is to have a view of the entire problem in order to define the needs in terms of numerical data and models. The Barnthouse model is interesting because its author is one of the few who have attempted to incorporate parameters of populations and ecosystems in the procedures of an ecological risk evaluation.

The second stage, *evaluation of the exposure–response relation*, quantifies the level of exposure and the level and probability of adverse effects. For the pollutants, this stage includes results of toxicity assays in the laboratory, epidemiological data, and concentrations of pollutants in the mediums, as well as appropriate models for analysing the data.

Next, *evaluation of exposure* establishes the level of exposure before and after implementation of regulatory measures.

Finally, *risk characterization* integrates the preceding information and communicates it to the manager, in a language comprehensible to non-specialists, generally in the form of a description of the nature and level of risk for elements at risk as well as qualitative and quantitative characterization of the uncertainty. Barnthouse emphasizes the difference between final point of measurement and final point of evaluation. The final points of evaluation are defined in terms appropriate to populations or ecosystems, for example, extinction of a population or changes in the structure of a community. The final points of evaluation are defined as a part of identification of danger. They are rarely measurable either because the evaluation is predictive (as with a new pesticide), or more generally because of the complexity and size of the receptor. For example, it is not possible to measure the effects of a pesticide on the entire population of birds in a region. The progress from final points of measurement to final points of evaluation requires extrapolation on the basis of models or expert judgement.

3.4.2. The Lipton Model

The Lipton model introduces original variations in comparison with the preceding models (Fig. 1.7; Lipton et al., 1993). Seven stages are described. Together they present interesting characteristics, notably a concern with incorporating ecological data, but the possibility of achieving all the stages in practice is doubtful.

Identification of receptors

Identification of receptors begins with a preliminary evaluation of distribution of stresses in the different compartments and the living elements (e.g., use of parameters such as $logK_{ow}$), then constructs simple models of the ecosystems affected, including their interactions, to characterize the structure of the system and the living elements present.

Identification of danger

The objective of identification of danger is the same as that of the corresponding stage of the NAS model—to determine whether exposure to a given agent may have adverse effects on health (and of what kind). However, in the case of ecological risk, the dangers must be identified for the receptors and levels of biological organization specified in the preceding stage.

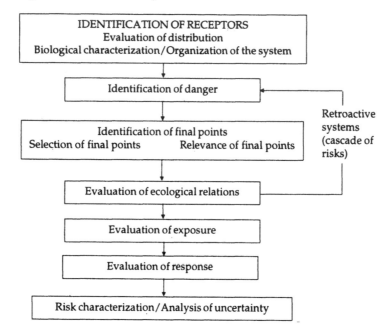

Fig. 1.7. General outline of ecological risk evaluation: the Lipton model (Lipton et al., 1993; with the permission of Springer-Verlag, 1997).

Identification of final points

Selection of final points is based on the receptors and dangers identified in the two preceding phases. It takes into account biological, economic, and socio-cultural elements. Ecological risk evaluations in response to regulatory demands are based on *standardized final points*, which ensure more coherent evaluations and eliminate some of the variation that exists when one identifies final points in natural systems. On the other hand, the evaluation may not take into account the real situation or the real components of biological systems.

Evaluation of ecological relations

Evaluation of ecological relations is the identification of relations between the receptors, retroactive links among the different living elements, if any, and eventual incorporation of secondary receptors into the process, then tertiary receptors (if necessary), which will be evaluated in their turn.

Evaluation of exposure and evaluation of response

These two stages are identical to those described in the NAS model (1983). The 'dose–response evaluation' of that model is here called 'evaluation of response', in order to account for the fact that many data are not accessible in the form of concentrations of the product and that certain stresses are not easily expressed in the form of dose (e.g., greenhouse effect, introduction of genetically modified organisms).

Risk characterization and analysis of uncertainty

The estimates of risk levels are calculated and presented with the associated degree of uncertainty.

3.5. Models of the PEC/PNEC Type

3.5.1. European Principles of Risk Evaluation

Rule 1488/94 of the European Union[26] establishes the principles of evaluation of risks for humans and the environment presented by existing substances. Several considerations of a general order have led the EU to set down such principles:

- The need for a system of risk evaluation.
- The need for principles common to all states.
- The fact that evaluation of risks to humans must take into account physiochemical and toxicological properties of a substance and evaluation of risks to the environment must take into account ecological events.
- The fact that decisions must be based on the results of a risk evaluation.

The Rule consequently established a certain number of definitions.

Identification of dangers is 'the identification of undesirable effects that a substance is inherently capable of causing'.

Evaluation of the relation of dose (concentration) to response (effect) is 'the estimation of the relation between the dose, or the level of exposure to a substance, and the incidence and gravity of an effect'.

Risk characterization is 'estimation of an incidence and gravity of undesirable effects likely to be produced in a human population or a component of the environment because of exposure, actual or foreseeable, to a substance; the characterization may comprise "risk estimation", that is, the quantification of this probability.'

Annexure III of this regulation specifies the general progress of an evaluation, based on the comparison of PEC and PNEC values. The PNEC is calculated from LD_{50}, NOEC, and other values, corrected by a factor of evaluation (cf. chapter 3, section 4.2).

---

[26]The principles of risk evaluation discussed in this paragraph are not specific to the EU but are perfectly representative of an approach of the PEC/PNEC type. Other examples of general models of evaluation can be found, based on approaches of the same type, in monograph no. 105 of the OCDE (OCDE, 1995).

Generally, in a PEC/PNEC approach, the PEC is calculated by a mathematical model incorporating and describing the occurrence and behaviour of products in a certain environment. The product of this model is a value or small series of values characterizing the maximum or mean environmental concentrations. The PNEC is a single value (LC_{50}, NOEC), for example for the most sensitive species, or the combination of values relative to a group of species.

3.5.2. Examples

The evaluation approach proposed by the ECETOC (1993) is an example of this practice. The objective is to obtain a risk evaluation by comparing the PEC values (obtained by exposure models) and PNEC values (based on toxicological data). Two geographic scales are studied, the local and the regional. At the regional scale, the mean environmental concentrations (diffuse pollution) are estimated. The local scale is studied more particularly for estimating concentrations of the pollutant near the source. The PNEC values are derived from acute or chronic toxicity data by use of factors of application[27] designed to compensate for differences between test conditions and natural conditions. The PEC is compared regularly to PNEC in a sequential evaluation, beginning with the regional scale. If the PNEC is greater than the PEC, the evaluation is terminated. Otherwise, the evaluation proceeds to the local scale and new, more detailed data are incorporated until the desired result is obtained. The model proposed by the ECETOC is not universal and the limits of its approach are specified:

- Metals are excluded and adaptation of the general model to petroleum hydrocarbons requires a supplementary operation.
- The proposed approach represents the state of the art at a given time, in a field that is rapidly evolving.
- The evaluation takes only freshwater organisms into account.
- The model presupposes a state of equilibrium, resulting from a continuous emission from the source. It is therefore not used for temporary emissions.

The possibility of incorporating field measurements (environmental concentrations or measurements of effects) is considered, but is not essential to the strategy.

Authorization procedures for new pesticides in the European Union countries are also based on a PEC/PNEC approach to evaluate the risk to aquatic and terrestrial fauna. PEC values are calculated for the different mediums (surface and underground waters, soil, and air) and PNEC values are defined (LC_{50} and NOEC for different animal and plant species). The

---

[27]The definition of factors of application (or factors of evaluation) and the principles used to calculate them are specified in chapter 3, section 4.2.

toxicity exposure ratios (TER = PNEC/PEC) are then calculated, and according to the value of these ratios, supplementary information is demanded or the product cannot be put on the market.

3.6. Models Applicable to Contaminated Sites

Several types of general models, such as the EPA model or Suter's retrospective model, may serve as conceptual frameworks for the evaluation of contaminated sites. However, specific models were developed for this purpose. Two examples, the EPA model (1989) and the Canadian model (1994), are given here.

3.6.1. The EPA Model

The EPA model (Warren-Hicks et al., 1989) was designed to be used in the preliminary phases of characterization of polluted sites, by a direct study of the site itself. The general approach is based on recognition of the fact that three types of information are necessary to establish a causal link between the presence of a polluted site and ecological effects:
- chemical analysis of polluted mediums to establish the nature, concentration, and variability of pollutant concentration;
- ecological investigations;
- toxicity assays to link environmental concentrations and toxic effects.

Without the combination of these relationships, the authors feel that it is not possible to eliminate the existence of factors of confusion such as the natural variability of living elements or the effect of natural disturbances (storms or floods, for example).

The objective is to obtain an evaluation of ecological effects, that is, at the level of populations and communities, from a combination of ecological final points that are considered 'reasonable' indicators of the state of these populations and communities. More precisely, the consequences of this evaluation are the following:
- an inventory of the present state of ecological final points at the site;
- an estimation of present adverse effects for which the polluted site is responsible;
- an estimation of the level and variation of adverse effects;
- to the extent possible, the attribution of these effects to pollutants of the site, rather than to natural factors or other disturbances of human origin.

The authors recognize that it is not possible to obtain a correct evaluation without taking into account the combination of chemical, ecological, and toxicological data, but that nevertheless, we are not sure of having completely eliminated these factors of confusion.

Some objectives are systematically excluded:
- prediction of future effects;
- an accurate evaluation of risk, even if the study data can be one component of a risk analysis;
- studies designed to optimize rehabilitation actions, evaluate the potential effects on human health, or evaluate the occurrence and behaviour of products present on the site;
- exhaustive ecological studies—the studies remain limited to the ecological final points defined earlier.

Conceptual models of ecological risk evaluation propose, in theory, to evaluate risk, that is, strictly speaking, to obtain a probabilistic evaluation of the level of risk. In fact, there is no well-documented example that shows that this level of evaluation has been attained, which is recognized explicitly by these authors, and therefore they call it an 'ecological assessment' rather than an 'ecological risk assessment'.

3.6.2. The Canadian Model

The Canadian model (Gaudet et al., 1994) is integrated in an overall strategy for management of polluted sites, the 'Programme National d'Assainissement des Lieux Contaminés', which has the ultimate objective of decision-making on the necessity or otherwise of rehabilitating a site. In most cases, the definition of *criteria of environmental quality*, based on pollutant levels that must not be surpassed in the environment, is sufficient. These criteria can be modulated according to particular considerations of the site. However, after a preliminary characterization of the site, it will sometimes be necessary to implement an ecological risk evaluation, especially if the application of criteria does not provide adequate protection and if important 'ecological components' are threatened. These supplementary 'triggers' are of three types (Table 1.1):
- significant ecological concerns;
- unacceptable lacunae in the data;
- particular characteristics of the site.

Process of evaluation

The different stages of ecological risk evaluation, as conceived by Gaudet et al. (1994), are the following:
- *identification of problem:* identification of key questions and objectives of protection;
- *characterization of places:* existing data on the geological, chemical, hydrological, and other characteristics of the place and its present and past uses;
- *evaluation of exposure:* sources of stress agents, scope, duration, and frequency of exposure;

Table 1.1. Supplementary triggers for an ecological risk evaluation in the Canadian model (according to Gaudet et al., 1994; with the permission of the Ministry of Public Works and Government Services of Canada, 1997)

Significant ecological concerns
- habitat threatened for fauna or flora
- populations or ecosystems threatened, presence of species in danger of or on the way to extinction
- natural reserves
- zones locally important for economic reasons: fishing, hunting, etc.

Unacceptable lacunae in the data
- pollutants with poorly understood toxic effects
- unpredictable or uncertain conditions of exposure
- strong uncertainty concerning the danger
- lacunae in information about ecological receptors

Particular characteristics of the site
- need to establish a priority if the cost of rehabilitation is high
- criteria existing before being tested in the field or improved
- polluted site too large for total rehabilitation

- *characterization of receptors:* identification of receptors and important habitats;
- *evaluation of danger:* characterization of ecological effects, toxicity of stress agents, modifying factors that can modulate toxicity in natural conditions, and measurement of responses;
- *characterization of risk:* biological response to concentrations of pollutants; scope, significance, and probability of the occurrence of effects as a function of the estimated exposure.

In this list, the habitats are explicitly mentioned among the receptors. An important step, which has not been made part of the evaluation objectives, is to plan an ultimate *biological surveillance* of the site.

The levels approach

An advantage of the Canadian approach is that it proposes a strategy by levels, with increasingly detailed evaluations. The authors suggest, in accordance with Maki and Duthie (1978), that a sequential evaluation, with retro-action, enables solid scientific judgements and rationalizes the collection of data, in the sense of an economy of means. They mention, however, that, according to Baker (1989), such an approach may multiply the field work, which is a source of additional cost and delays.

The information necessary for the three levels is presented in Figure 1.8. Overall, level 1 is characterized by simple qualitative or quantitative methods, based essentially on existing data. Level 2 provides semi-quantitative information from standard models and specific approaches. Level 3, the most complex, is based on specific data and predictive modelling that provides quantitative data.

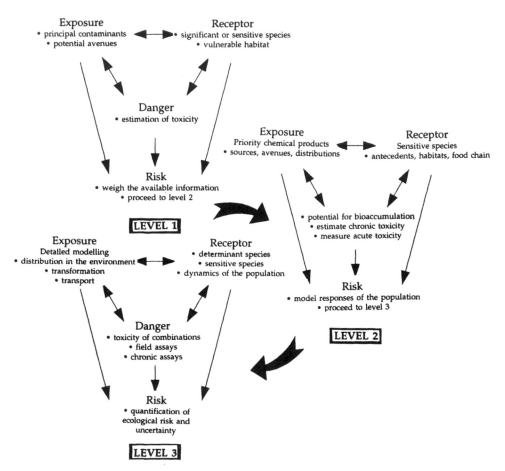

Fig. 1.8. The three levels of the Canadian model (Gaudet et al., 1994, with the permission of the Ministry of Public Works and Government Services of Canada, 1997).

2

Characterization of Exposure

1. Introduction

1.1. What is Exposure?

Exposure of living beings refers to 'contact with a [physical or] chemical agent' (EPA, 1986). The evaluation of exposure has been defined as 'measurement and estimation of the intensity, duration, and frequency of exposure to a dangerous product' (National Research Council, 1983). For Suter (1993a), it is 'an operation that links source to effect'. The following must be precisely defined before such an evaluation can be made (cf. chapter 1, section 2.2):
- spatial and temporal scope;
- elements present (X the pollutant and Y the polluted);
- forces that drive them ($\rightarrow$);
- their interaction (XY).

1.2. Where Does Exposure Begin and End?

If the preceding definition is taken literally, characterization of exposure aims only to describe the interaction between the pollutant and the polluted, that is, to describe what happens from the instant of contact between the two. However, the dividing line between the exposure phase and other phases varies according to the conceptual model. In the model of Covello and Merkhofer (1993), evaluation of sources and evaluation of exposure are two distinct stages of the evaluation process. In the models of Suter (1993a), the description of the environment and that of sources are separate and the exposure phase does not include the occurrence of the product in the environment. Finally, in the EPA model (Norton et al., 1992), characterization of exposure includes not only interaction, but also the circumstances that have made it possible, such as emission by sources, the occurrence of pollutants in the environment, and the distribution of exposed populations.

It is difficult to define the moment at which exposure ends and toxic effects begin. For a chemical product present in the environment to be toxic to a living being, it must come into contact with the physical limits of the organism, cross these limits, penetrate the organism, then make a stable chemical liaison with a molecular target. When this target plays an important role in the maintenance of life processes, the pollutant causes cellular dysfunctions that lead to the toxic effects that gradually overcome the tissues and the organism. There are thus two successive exposures, first the contact with the physical limits of the organism, then the contact with the molecular target. In most cases, the contact is with the physical limits of the living being, whether that be a bacterium, a bird, a fish, or a human, and the exposure is based on environmental concentrations of pollutants in the mediums (external doses). But in certain *multi-media* models (cf. section 4.4.2), the biomass is treated as a particular compartment with a defined volume and measurable capacity of exchange with other compartments. These models directly calculate internal doses. Finally, there is a particular category of models, the *toxicokinetic models* (or pharmacokinetic, because derived from studies on the effects of drugs), that describe the occurrence of a product in the organism and evaluate doses at the level of the molecular targets. These models, sometimes used for human risk evaluation, are classified as either exposure scenarios or effect scenarios (Covello and Merkhofer, 1993). They are still poorly developed for species other than humans.

1.3. The Particular Case of Ecological Risk

Evaluation of exposure is more complex in the case of an ecological risk than in the case of a health risk because:
- Several different species are exposed.
- *Direct* exposure to a product via a contaminated food or medium causes *direct* toxic effects on these species, but also *indirect* effects on other species, for example by competition for sources of food or habitats. This possibility is not evaluated, properly speaking, in the exposure phase, but it is essential to foresee it in the exposure scenario (cf. section 4.3.1).
- The evaluation must go beyond the level of the individual and take populations and ecosystems into account.

Most exposure scenarios developed for human risk evaluation can be used for ecological risk, at least partly. For example, the data and models relative to the occurrence and behaviour of the pollutant in the soil and to its transfer in plants can be used in evaluation of risk to birds and herbivorous mammals.

1.4. The Exposure Scenario

The exposure scenario is one part of the general scenario. It describes in detail the different elements constituting the exposure phase:

- the mediums;
- the sources: description and quantification of the emission (nature of pollutants emitted, intensity, frequency, and duration of emissions);
- the spatio-temporal evolution of products emitted in the environment: avenues of transfer and occurrence of product;
- the spatio-temporal evolution of populations at risk in this environment, in the absence and presence of products (attraction/repulsion reactions);
- the nature of points of contact with the products (avenues of exposure);
- the extent of contact: intensity, duration, and frequency of exposure (doses at point of contact, external doses) of targets (organisms);
- the internal doses (contamination of targets) evaluated from the relative significance of different transfers (bioavailability);
- the means of transfer and distribution in the organisms towards the molecular targets;
- the (critical) doses reacting with the molecular targets.

The last two points pertain only to toxicokinetic models. The final presentation of the exposure scenario includes:

- a summary of conclusions of the risk formulation phase, identifying the dangerous products (stresses) and the elements at risk (receptors, final points of evaluation), which serves as a base from which to develop the scenario;
- operational tools such as hypotheses, qualitative models (diagrams of avenues of exposure), quantitative models (mathematical models), and factual data generated by the different approaches (specific experiments at the site or compilation of data existing in the literature);
- the final description of results obtained: avenues of exposure, intensity, duration, and frequency of exposure of elements at risk.

The methods are those described in chapter 1, conventional tests, bioassays, mathematical models, and *in situ* indicators. In this chapter, we present those that are the most appropriate for characterization of polluted sites.

1.5. Polluted Sites: The Canadian Approach

The Canadian approach is based essentially on that developed in the United States by the Environmental Protection Agency and summarized in several important documents, such as the *Superfund Exposure Assessment*

Manual (EPA, 1988), the *Superfund Human Health Evaluation Manual* (EPA, 1989a), which replaced the *Superfund Public Health Assessment Manual* (EPA, 1985), and *Risk Assessment Guidance for Superfund Sites* (EPA, 1989b).

Evaluation of exposure, according to the EPA, attempts to answer the following questions (in Gaudet et al., 1994):

- What are the organisms that are or may be exposed to dangerous pollutants?
- What are the significant avenues of exposure?
- What are the doses, duration, and frequency of exposure?
- What are the seasonal variations that can affect exposure?
- What are the geophysical, physical, and chemical characteristics specific to the place and capable of influencing the exposure?

These questions have not been included in the 'exposure' stage of the Canadian approach (cf. chapter 1, Fig. 1.8), since the identification of receptors is the object of a particular stage. In fact, receptors must be identified earlier, during risk formulation, but it is evident that the more exhaustive characterization of pollutants present in a site can bring about a modification in the initial list of receptors, for example, if populations or communities other than those that were identified at first were found to be at risk of exposure. This is one reason flexibility and iterative nature are essential to risk evaluation studies.

The key elements of the exposure, according to the Canadian approach, are summarized in Table 2.1. One of the most delicate points is the choice of pollutants to be incorporated in the evaluation, for which there do not seem to be universally established criteria. Three general principles have been proposed by the EPA (1989b, cited by Gaudet et al., 1994):

- identification of physical and chemical properties of pollutants present on the site;
- the grouping of pollutions according to three criteria: their physical and chemical properties, the medium, and the living elements present in it.
- selection of the most toxic pollutant in each group, according to the available information on the concentrations measured and the dose–response relations.

2. CHARACTERIZATION OF ENVIRONMENT AND SOURCES

The following is a summary and specification of concepts already introduced. Many techniques are available: for further details, the reader is invited to refer to the numerous works that exist on the subject.

The *scale* of an evaluation is a very important parameter to take into account. An evaluation on the *local* scale is not done in the same way as

Table 2.1. Key elements of exposure (according to Gaudet et al., 1994, with the permission of Ministry of Public Works and Government Services of Canada, 1997)

- Evaluation of target chemical products
- Release of pollutants
- Analysis of transfer and occurrence of contaminants
- Analysis of avenues of exposure
- Quantification of exposure of aquatic receptors
- Quantification of exposure of terrestrial receptors
- Analysis of uncertainty

one on the *regional* scale, the local scale representing some metres to a few kilometres, and the regional scale covering hundreds or thousands of kilometres (Sheehan, 1995).[1] In terms of terrestrial pollution, these two scales correspond quite well to the two situations described at the beginning of this volume, polluted sites (local scale) and polluted soils (regional scale). The final points of measurement and evaluation are not the same on the local and regional scales. Examples can be found in Sheehan (1995). On regional-scale risk evaluation, the reader can also consult Hunsaker et al. (1990).

The *modalities of emission* are estimates of spatio-temporal characteristics of the introduction of the pollutant in the system, from a source or a group of sources. The precision of these estimates varies widely. Although the volume of liquid effluents or incinerator fumes, can be measured with some precision, we have a less precise idea of contents of a particular pollutant in the incinerator fumes, and an even less precise idea of the distribution of mutagenic products. The significance of these emissions can be estimated by statistics, for example, the volume of pesticides sold or quantities of products discharged (with all the imprecision involved in such estimations). Thorough researches with local or national administrative bodies (commons, cities, departments) and previous owners will be useful in determining the nature and quantities of products manufactured, stored, or dispersed on the site.

Simulation techniques are helpful, for example, to calculate the possibility of a canal transporting a toxic product or the rupture of a shaft in a soil or an abandoned mine. These different models of chemical accidents have been described in detail by Covello and Merkhofer (1993).

The distinction between *source* and *medium* at risk is easy to establish in the case of localized industrial pollutions, such as an incinerator or factory effluent. The case of polluted sites and soils is different: a very polluted soil, contaminating the air and water and nearby land, is comparable

---

[1]The OCDE (1993c) defines *local* models (of exposure), which describe the occurrence and behaviour of products near sources, and *global* models, which describe the occurrence and behaviour of products on the global, continental, regional, and national scale.

to a source of pollution, whereas vast areas of land in contact with a more diffuse pollution, the effects of which are not readily apparent, are considered ecosystems at risk (cf. chapter 1, section 3.3.2).

In the particular case of polluted sites, preliminary investigations help us make an approximate delimitation. The primary information on the extent and significance of the pollution often comes from historical data, such as the recognized existence of a site of discharge, an old factory, or a disused mine.

The abiotic medium must also be characterized by its physico-chemical characteristics. Many field data already exist, such as hydrological, geographical, geological, and pedological maps and meteorological data, but some can be generated during the study (physico-chemical analysis of the soil, for example).

This primary information is confirmed by measurements of pollutants in the soil, possibly in the nearby water courses, the phreatic layer, and the atmosphere. The sampling strategies and techniques of sampling, and the completion of physico-chemical analysis are developed in detail in specialized works and are not summarized here. The analytical data can be used with software that enables the tracing of isodensity curves of pollutants. The collection of reliable analytical data is of definite importance, but whatever the quality of analysis, the pollution of soils, especially of polluted sites, is fundamentally heterogeneous, which seriously complicates their evaluation.

The *limits of the evaluation* are not those of the polluted site, but those of the ecosystem at risk, that is, all the area susceptible to being polluted by this site. For example, to evaluate the ecological risk of a polluted site situated on the banks of a sloped basin, it is essential to pay particular attention to the areas downstream of the site, the risk being less for the animals and plants upstream of the site. In a retrospective evaluation, we can define the limits according to the values measured, for example, the dangerous area is that in which the pollutant content (or the toxic effects determined by bioassays) is higher than a limit fixed beforehand. This method was used by Thomas et al. (1986) for the evaluation of the site of an arsenal in the Rocky Mountains (USA). Samples of polluted soil were put through a series of bioassays and graduated responses thus obtained enabled the charting of a map of pollution levels on the site.

Other limits are currently used, even though they are not scientifically justified, for example, administrative limits (departments or regions).

Finally, some authors have proposed the use of Geographic Information Systems data for the purpose of forecasting (Stallones et al., 1992).

3. CHARACTERIZATION OF OCCURRENCE AND BEHAVIOUR IN ENVIRONMENTS

The occurrence and behaviour of products in environments can be described at three levels (OCDE, 1993a):

- the environment in its entirety;
- adjacent parts of the environment;
- a single part of the environment.

The following discussion essentially pertains to soil. Soil is a complex system, composed of a liquid phase, a solid phase, and a gaseous phase in interface with the atmosphere, lithosphere, and hydrosphere. A soil is defined by its *texture* (relative proportion of different mineral and organic components, such as sand, lime, clay, clay-humus complex, microflora, and microfauna) and its *structure*, which represents the spatial organization of these components. The pollutant in the polluted site migrates in the soil and distributes itself in the different phases that constitute the soil, becomes volatile in the atmosphere, and contaminates water courses. The movement of products in soils is determined by two major phenomena: (1) exchanges between the immobile solid phase and the mobile gaseous and aqueous phases and (2) movements of the mobile phases. The combination of these phenomena determines the *persistence* of products in the soil.

The transfer does not occur only between abiotic mediums. Transfer to biomass (plants and animals) is a particular type of exchange that will be discussed later.

The occurrence and behaviour of pollutants in soils have been reviewed in detail (I2C2, 1994). The summaries that follow, designed to enable a continuous discussion, concern pollutants in a polluted site that is considered as a source of pollution.

3.1. Physico-chemical Characteristics of Soils

Several physico-chemical characteristics of soils are important in explaining and forecasting the behaviour of pollutants:

- porosity (S_{np}, $m^3 \cdot m^{-3}$);
- content in air and water (S_{na} and S_{nw}, $m^3 \cdot m^{-3}$);
- density (S_g, $g \cdot m^{-3}$);
- permeability;
- granulometric composition: distribution of particulate dimensions from three categories—clay, lime, and sand—defined respectively as particles measuring less than 2 μ, 2–50 μ, and 50–2000 μ;
- content of organic carbon (f_{oc});
- cation exchange capacity;

- pH;
- redox potential.

Additional information can be found in specialized texts, such as that of Baize and Jabiol (1995).

3.2. Soil–Water Exchanges

3.2.1. Phenomena at Work

Soil–water exchanges are essentially:
- exchanges between the solid phase and liquid phase (by diffusion);
- displacements of the soil solution by convection (leaching, lixiviation, percolation[2]);
- surface effects (flows).

Exchanges between the solid and liquid phases of soil are characterized by absorption–desorption coefficients that express the equilibrium between the retention of pollutant in the solid phase and solubility in the aqueous phase. The quantity of pollutant present in the soil solution determines the mobile fraction. The bioavailable fraction, which is the fraction of pollutant penetrating a living element (plant or earthworm), depends on the mobile fraction and characteristics of the living element. The term *bioavailable* is also applied to the fraction of the product that penetrates an organism in relation to the fraction absorbed.

For organic pollutants, the phenomena of adsorption are described by isotherms of adsorption, the most current model of which is the isotherm of Freundlich:

$$(x/m) = K_f \times C_e^n$$

where x/m is the quantity of pollutant adsorbed in the soil, C_e is the concentration of pollutant in the aqueous solution in equilibrium with the adsorbed phase, K_f is the capacity of adsorption, and n is an index of affinity of the pollutant for the soil. For many molecules, n is close to 1, from which we derive the simplified equation:

$$K_d = (x/m)/C_e$$

The coefficient of distribution K_d is generally normalized by being related to the organic carbon content of soil (K_{oc}):

$$K_{oc} = K_d \times 100/\% \text{ carbon}$$

---

[2]Leaching: forcing of particles into suspension by convection. Lixiviation: transport of dissolved substances in the soil by convection. Percolation: vertical transfer of a liquid in a porous medium.

One difficulty in studying these phenomena is the temporal dimension. The absorption–desorption constants are equilibrium values that do not provide any indication of the time needed to reach this equilibrium. Moreover, this description of phenomena is useful only for relatively low pollutant concentrations, which do not disturb the physico-chemical characteristics of the soil or the microbial fauna and flora. Very high concentrations, in solvents for example, introduce toxic liquid phases into the soil that are not miscible with water, the behaviour of which cannot be analysed with conventional models. Finally, soils adapt to the presence of pollutants. The rate of degradation increases considerably in soils polluted at regular intervals, by selection of colonies of adapted bacteria.

3.2.2. Methods of Evaluating Mobility in Aqueous Phase

Table 2.2 summarizes different methods that can be used to estimate the mobility of pollutants in soil. The physical models vary almost infinitely in size and complexity. The most simple are laboratory tests. Determination of the octanol–water partition coefficient is a typical example of an extremely simplified model, composed of two phases and a pollutant, enabling us to characterize the capacity for transfer between an organic phase and an aqueous phase and to calculate the concentrations at equilibrium. The K_f or K_d values can be calculated experimentally for a given molecule and soil with the standard protocols presently available. These values can also be estimated from other parameters. As the adsorption on soils reflects the distribution between an aqueous phase and an organic phase (soil organic matter), it is conceivable that the parameters representative of partition potentialities between these two phases, such as solubility in water or the octanol–water partition coefficient, are indicators of adsorption.

Correlations between K_{oc}, S_w (solubility in water), and K_d have been established for many classes of products. Table 2.3 and Figure 2.1 show some examples. The adsorption–desorption coefficients can also be estimated from molecular characteristics, such as the Hansch constant or connectivity indices (Sabljic and Piver, 1992). Other relations of this type can be found in OCDE monograph no. 67 (OCDE, 1993a).

The progression to more complex models and a larger spatial area has the advantage of highlighting phenomena inaccessible to reduced scales of investigation. The rate of percolation in a soil depends on adsorption–desorption characteristics of the product, and on physical characteristics (existence of cracks, etc.) found in scales difficult to reproduce in a laboratory. These phenomena will not be apparent except by experiments in lysimetric cases or in the field. However, while the most complex physical models ensure a better description of phenomena in a given situation, they require difficult experiments and cannot be analysed easily by present mathematical models.

Table 2.2. Methods of study of mobility and persistence (according to I2C2, 1994)

Mobility

Organic pollutants
- measurement of $\log K_{ow}$
- measurement of adsorption coefficients (K_d) and standard adsorption coefficient (K_{oc})
- measurement of desorption in laboratory and laboratory extraction techniques (pressure, centrifuge, capillarity)
- soil columns, thin layers of soil
- lysimetric cases
- porous candles *in situ*
- soil monoliths

Trace elements
- partial chemical extractions

Persistence (biodegradation)

Organic pollutants
- tests of biodegradation by microflora and microfauna (laboratory methods)

Trace elements
evolution of pH and redox conditions

Transfer in plants

Organic pollutants
- tests of degradation (the bioavailable fraction corresponds to the biodegraded fraction)
- toxicity tests (tests of germination and root growth)

Trace elements
- chemical extraction
- bioassays
- analysis of spontaneous vegetation

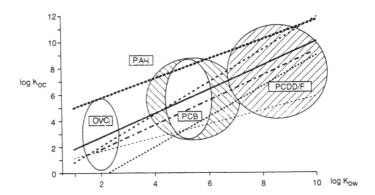

Fig. 2.1. Estimates of K_{oc} from K_{ow} values according to equations from Table 2.3 and delimitation of margins of variation of K_{ow} for different organic pollutants (OVC: organo-volatile compounds; according to I2C2, 1994; with the permission of RECORD, 1997)

Table 2.3. Examples of estimation of K_{oc} value from values of solubility in water (S_w) for different organic compounds (according to I2C2, 1994)

Authors	Linear correlation	Type of compound
Hassett et al., 1983	$LogK_{oc} = 3.95-0.62 \, logS_w$	(Various non-polar compounds)
Chiou et al., 1979	$LogK_{oc} = 4.04-0.56 \, logS_w$	(CI-aliphatic and aromatic)
Karichof, 1981	$LogK_{oc} = -1.4-0.92 \, logS_w-$ $0.0095 \, (Mp^*)$	(Aliphatic and aromatic) $(^*Mp = temperature \, of \, fusion)$
Gerstl and Mingelgrin, 1984	$LogK_{oc} = 3.80-0.56 \, logS_w$	(Various pesticides)
Moreale and van Bladel, 1982	$LogK_{oc} = 2.75-0.45 \, logK_{ow}$	(Various compounds)
Hassett et al., 1983	$LogK_{oc} = 0.88 + 0.91 \, logK_{ow}$	(Various non-polar compounds)
Chiou et al., 1983	$LogK_{oc} = 0.78 + 0.90 \, logK_{ow}$	CI-aromatic
Vowles and Mantoura, 1984	$LogK_{oc} = -2.53 + 1.15 \, logK_{ow}$	Aromatic hydrocarbons
Gerstl and Mingelgrin, 1984	$LogK_{oc} = 4.4 + 0.72 \, logK_{ow}$	Various pesticides
Madhum et al., 1986	$LogK_{oc} = -0.40 + 1.23 \, logK_{ow}$	Various pesticides
Brown and Flag, 1981	$LogK_{oc} = 0.01 + 0.94 \, logK_{ow}$	Various pesticides
Briggs, 1981	$LogK_{oc} = 0.69 + 0.52 \, logK_{ow}$	Various pesticides

The phenomena that will be described are not independent of each other. There are relationships between mobility, degradation, and other phenomena that can be integrated in the most complex models, but they are often incorporated in exposure scenarios in the form of default values or correction factors (additive or multiplicative) or are simply ignored.

The evaluation of capacities of degradation in soils, enabling us to calculate persistence in the form of half-life ($t_{1/2}$), is particularly difficult to establish. An evaluation of risk of pollution can be obtained with models such as the empirical index GUS (groundwater ubiquity score) proposed by Gustafson (in I2C2, 1994):

$$GUS = \log(t_{1/2}) \times (4 - \log K_{oc})$$

Pesticides having GUS values less than 1.8 will have less risk of reappearing in the water of phreatic layers, whereas the most frequent cases of pollution of the phreatic layer correspond to pesticides having GUS values greater than 2.8.

The occurrence in soils and the possibilities of reaching phreatic layers or surface waters depend on the precipitation regime. The climatic conditions on which the evaluation is based will depend on the spatial scale envisaged-polluted site or diffuse pollution in a country or continent. The values used could be mean values (annual or monthly) or extreme values, depending on the available data and the objectives of the study. The most simple models take only precipitation volumes into account, and the most elaborate take precipitation regimes into account.

The pollutants capable of migrating by flow lie in the top centimetre of the soil. The significance of the flow will depend on the force of the precipitations.

3.3. Soil–Air Exchanges

There is a deposit of atmospheric particles from the air to the soil, and diffusion towards the soil. From the soil to the air, the principal phenomena are diffusion in gaseous phase, volatilization from the surface, and soil erosion. These phenomena are complex, but the volume of transfers of pollutants by volatilization, even for molecules with very low vapour tensions, must be taken into account. Once in the atmosphere, the pollutants will be adsorbed/desorbed in atmospheric dust, dissolved in water drops in clouds, and transported by wind and clouds. During these transfers, the products undergo molecular transformations by photo-oxidation or hydrolysis, degradation reactions, or, conversely, formation of chemically reactive entities.

Volatilization has been particularly studied in water. For example, the rate of volatilization can be estimated by the Liss-Slater model, which allows us to calculate the speed of volatilization from the water to the air (k_v) from different physico-chemical parameters (e.g., Henry constant,[3] constants of molecular diffusion of the air and of the pollutant in water and of the pollutant and the water in air, depth of the water layer). This leads to equations of the following type (Mill, 1993):

(1) $k_v = H_c k^w_g D^r_g / L$ ($H_c < 1000$)

or:

(2) $k_v = k^o_i (D^r_i) \cdot 0.7/L$ ($H_c > 1000$)

where k_v is the constant of speed of volatilization, H_c is the Henry constant, k^w_g is the constant of diffusion of water vapour in air, D^r_g is the ratio of constants of molecular diffusion of the product and water in air, L is the depth of the water layer in cm, k^o_i is the coefficient of transfer of mass, and D^r_i is the ratio of constants of molecular diffusion of the product and the air in water.

Volatilization from soil is a very complex phenomenon. It would seem that while the content of water is high, the transfer occurs by the intermediary of an aqueous phase, whereas in dry soils, there is a direct transfer

---

[3]The Henry constant describes the distribution of a substance between air and water. It is expressed either as the ratio of partial pressure of the gaseous phase to the concentration in water ($H = P/C_w$), or as the ratio of concentrations in the air and water ($H' - C_a/C_w$).

from the solid phase to the gaseous phase. At present, there are no simple laboratory models that take this phenomenon into account. In the initial phase, volatilization from the superficial aqueous phase will be the limiting stage and, according to Mill (1993), it is possible to use an equation of type (2), but the limits of this model are very rapidly reached because of the development of a concentration gradient along the soil column. A detailed review of volatilization of pesticides from soil can be found in an article by Spencer et al. (1973).

3.4. Transformation Reactions

The capacity of degradation of soils can be viewed in two ways, either as a factor modulating the exposure of receptors such as humans, terrestrial vertebrates, or underground species, or as a loss of the functional capacity of the soil ecosystem.

Degradation of a pollutant is one form of mobility, because it leads to its disappearance in the soil. Transformation reactions of pollutants in the soil are due to biotic and/or abiotic reactions. Trace elements undergo changes in the chemical state (speciation), such as oxidation-reduction, or formation of more or less soluble salts or organometallic complexes. The occurrence and evolution of these different chemical forms of the same elements in the mediums are often very different, as is their toxicity. It is well known that mercury and methyl mercury do not have the same physico-chemical properties or the same toxicity.

Organic pollutants are subject to hydrolysis or oxidation–reduction reactions that lead to the ultimate degradation of carbon chains into CO_2. These molecular transformations are catalysed by biotic and abiotic reactions according to the conditions of the medium (pH, redox potential, temperature, presence of a particular microflora and microfauna). The constants of speed of hydrolysis reactions in water are measured in the laboratory by standard methods. The biological activity appears mostly on the soil surface, but there are examples of degradation of certain pesticides at depth.

OCDE monograph no. 68 (1993b) presents a detailed review of biodegradation and the corresponding evaluation methods.

3.5. Occurrence and Behaviour on the Scale of Ecosystems

The preceding descriptions apply to exchanges between two mediums, soil and water or soil and air. More recently, the analysis of transfer phenomena was transposed to the scale of ecosystems, with the development of multi-media models comprising all the phenomena of transfer between the different abiotic and biotic mediums (definition of a 'biomass' compartment). Examples of these models can be found in Annexure 1.

3.6. Interpretation of Ground Data

The concentration of pollutants in the different mediums can be measured in the polluted site itself (soil) and near the site (atmosphere, water course crossing the polluted site, or the ecosystem at risk). These data serve as parameters and are incorporated in the mathematical models. They have two types of uses:
- to characterize the exposure;
- if there is a prompt follow-up, to construct statistical models, analyse trends, and anticipate the overall development.

4. CHARACTERIZATION OF EXPOSURE IN BIOLOGICAL SYSTEMS

Once environmental concentrations are determined, with some precision in localization and values, the exposure of living elements must be evaluated. This phase of risk evaluation is one of the most difficult, and the modelling of exposure of living elements is often summary. In most cases, we settle for a simplified description of the exposed populations and possible avenues of exposure. For example, to evaluate the exposure of water birds to a polluted sea, we calculate the exposure of an 'average' bird considered representative of individuals making up the different bird populations that may be present on the site. The levels of exposure are based on *mean* values (the average weight of an adult, the average daily consumption of water), and calculated by algorithms such as those presented in the following paragraphs or in Annexure 1. In a worst-case strategy, *conservative* values are used (for example, the most sensitive species, individual, or stage of development) in place of mean values.

4.1. Exposure of the Soil Ecosystem

In the case of a polluted soil, the species underground and on the surface are exposed to pollutants by:
- direct contact with the contaminated mediums;
- indirect contact, through ingestion of contaminated soil or food (plants or animals).

These different avenues of exposure are difficult to differentiate. For the microflora and microfauna, we do not seek to characterize the exposure of a species very precisely, only to identify the relationship between the environmental concentration and parameters considered representative of proper functioning of the soil ecosystem. The ecotoxicity assays for soil fauna (e.g. earthworms, Collembola) establish the relation between concentrations in the soil and various toxic effects (generally the LD_{50}), without the need to know the avenues of exposure (contact with the soil solution and/or ingestion of soil) and the bioavailability of pollutants. However, a more precise knowledge of these avenues of exposure, the distribution of pollutants between the different phases, and variations in sensitivity to toxins that result from it will facilitate the extrapolation of results to local conditions, which may be diverse.

4.2. Transfer into Plants

Transfer into plants has two consequences:
- the appearance of phytotoxic effects (themselves a source of modifications of food resources for herbivores) and
- the presence of potentially toxic pollutants in the food of herbivores.

Transfer into plants, from soil, occurs by root absorption or by gaseous diffusion. A third possible means is the deposit of atmospheric particles of local or distant origin. One of the best methods for estimating the possibilities of transfer into plants consists of cultivating plants on polluted soil, under defined experimental conditions, and measuring the pollutants in various organs (*bioassays*).

The literature on the subject shows that plants, like animals, have capacities of storage or, conversely, degradation, that vary widely according to species, nature of the pollutant, and antagonism and synergy between toxic molecules. Plants can bioconcentrate lipophilic organic contaminants. Several authors (Travis and Hattemer-Frey, 1988; Bacci et al., 1990a, b) have shown that bioconcentration in plants follows a relation of the following type:

$$logBCF = a + b \cdot logK_{ow}$$

where BCF is the bioconcentration factor (cf. section 4.3.3). Plant species with high capacity for bioconcentration can extract pollutants and decontaminate a site. Their relatively small biomass, however, limits the quantities of pollutant that can be extracted.

In the case of organic pollutants, root absorption is a passive process, depending on the polarity of molecules. The translocation from the roots to above-ground parts is correlated with K_{ow} values, and according to Briggs et al. (1982), there is an optimal value ($logK_{ow} = 1.8$) for a maximum translocation towards the above-ground parts. The highly lipophilic pollutants ($logK_{ow} > 6$), such as PCDD/PCDF, remain adsorbed on the roots with very poor mobilization toward the interior of the plant. However, the $logK_{ow}$ is not enough to explain the interspecific variations of the rate of transfer in plants, for example, certain representatives of the Cucurbitaceae accumulate significant quantities of PCDD/PCDF (Hulster et al., 1994).

The contamination of aerial parts of plants by gaseous modes is an avenue that must not be neglected, above all in the case of hydrophobic organic pollutants such as PAH[4] (Simonich and Hites, 1994).

Some mathematical models have been developed to evaluate transfers from soil to leaves. Figure 2.2 represents the model proposed by Taylor et al. (1988). This model is based on the resistance of different plant elements to gaseous movements and works on the basis of Ohm's law:

$$J = C/R$$

where J is the rate of deposit, C is the gradient of pollutant (the 'difference of potential'), and R is the resistance of various elements.

A mathematical model has been developed by Boersma et al. (1988 and 1991) to study the transfer of bromacil (a herbicide) in soy. This model is based on a 'standard' plant, composed of a root, three stems, and three leaves, each of these compartments being divided into three sub-compartments (phloem, xylem, and storage). Several parameters are taken into account: photon flow, air temperature, relative humidity (3 possible values), wind velocity, CO_2 content, evapotranspiration (3 possible values), and bromacil concentration. There are models of transfer toward aquatic macrophytes (Gobas et al., 1991). Descriptions of other models can be found in a review by Kabata-Pendias and Pendias (1984). Structure–activity relationships designed to predict transfer toward plants can be found in the articles of Topp et al. (1986) and Travis and Arms (1988).

4.3. Exposure of Terrestrial and Aquatic Fauna: Predictive Estimates

4.3.1. Trophic Networks

Figure 2.3 represents the principal groups in which living elements in the biosphere are distributed: primary producers (plants, algae), primary consumers (herbivores), secondary and tertiary consumers (carnivores), and decomposers. Another more detailed example is presented in chapter 3, Tables 3.3 and 3.4. The trophic chain is not as linear as this classification makes it seem (the carnivores consume the herbivores that consume the plants): higher levels may have several sources of food. For example, humans (omnivores), in addition to a direct contact with the mediums (air, soil, water), may be exposed to toxins by consumption of plants and (generally herbivorous) animals. A theoretic example of exposure in

---

[4]PAH = polycyclic aromatic hydrocarbons.

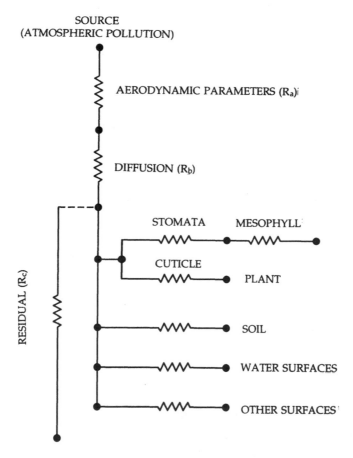

Fig. 2.2. Model of transfer of gaseous pollutants towards plants (the different 'resistance', according to Taylor et al., 1988; with the permission of Lewis Publishers (CRC Press), 1997).

terrestrial and aquatic food chains is found in Gaudet et al. (1994) (Figs. 2.4, 2.5). Another example, corresponding to an actual case, is given by MacIntosh et al. (1994) (Fig. 2.6).

It must be noted that each link in the trophic chain is either an element at risk or simply a risk factor in the trophic chain of another element at risk. Plants, for example, are elements at risk if the objective of the evaluation is to measure the loss of plant biodiversity. At the same time, the presence of toxic residues in plants is a direct risk of toxicity for the primary consumers (and of secondary toxicity for secondary consumers; cf. section 1.3), while the disappearance of a plant is in itself an indirect risk for insects depending on that plant. In another example, rat poison presents a high risk of *direct toxicity* for rodents, which are a population at risk, while presenting also a risk of secondary toxicity by consumption

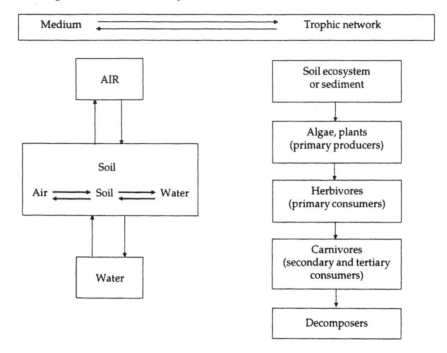

Fig. 2.3. Exposure of living elements of an ecosystem to pollutants according to their trophic rank.

of rodents containing significant quantities of rat poison, and finally an indirect risk by seriously reducing the population of rodents, which are an important source of food for birds of prey.[5]

In abiotic mediums, concentrations of pollutants are not identical in different links of the trophic chain, because of varying exposures to the polluted mediums and variable capacity to store or metabolize the pollutants. The phenomena of bioconcentration and bioaccumulation, because of the additional risk they represent, are discussed below.

4.3.2. Avenues and Models of Exposure

Aquatic species

The determination of avenues of exposure is not important for *fish* exposed to water-soluble pollutants and poor in bioaccumulation, because toxicity

---

[5]DeSnoo et al. (1994) also distinguished between direct and indirect unintentional effects (side effects). Direct effects are *primary* when the toxin has an adverse effect on a non-target organism (e.g., pesticides), and *secondary* when the toxin is transferred in the trophic chain. Indirect effects are effects that are not due to direct action of the toxin, for example, the disappearance of prey or vegetation. However, the notion of direct, indirect, and secondary effects is not always defined in the same manner in the ecotoxicology literature.

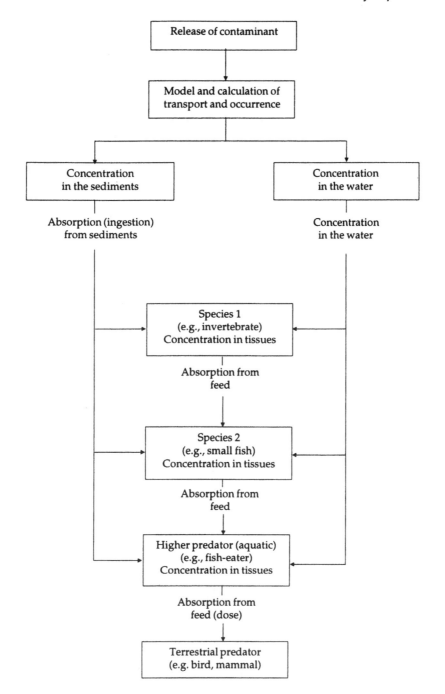

Fig. 2.4. Exposure in aquatic food chains (according to Gaudet et al., 1994; with the permission of the Ministry of Public Works and Government Services of Canada, 1997).

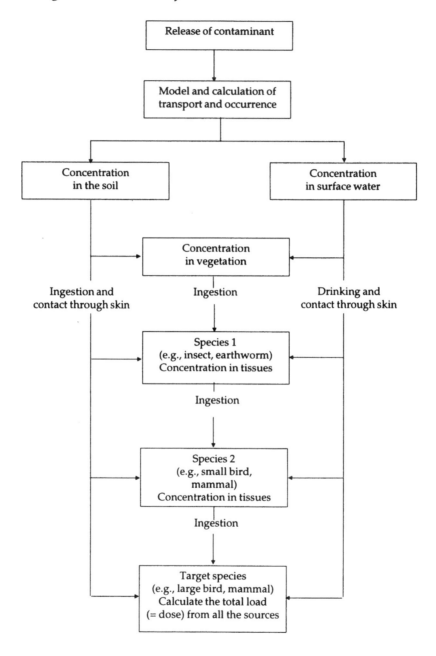

Fig. 2.5. Exposure in terrestrial food chains (according to Gaudet et al., 1994; with the permission of the Ministry of Public Works and Government Services of Canada, 1997).

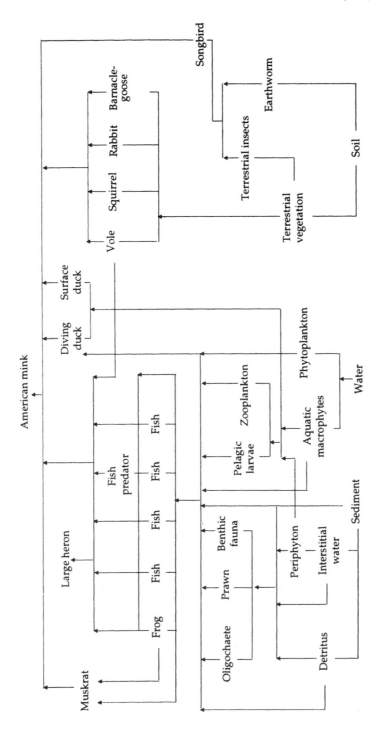

Fig. 2.6. Example of avenues of exposure of two species at risk, heron and mink (according to MacIntosh et al., 1994; with the permission of Plenum Press, 1997)

tests reproduce these conditions easily enough. In the case of pollutants that are poorly soluble and strongly adsorbed on the *sediments*, the possibility of contamination by ingestion of contaminated particles, plant debris, or something else complicates the analysis of the problem. It is therefore important to know whether the predominant avenue of exposure is contact with the aqueous phase and penetration by way of gills, or ingestion of contaminated particles.

Some models of exposure of aquatic fauna, essentially fish, have been developed in the past few years, for example, the FGETS model (food and gill exchange of toxic substance), devised to predict the bioaccumulation of non-polar organic pollutants in fish. Other examples are found in Annexure 1 (multi-media models) and section 4.3.3.

The interaction of benthic organisms with a pollutant being adsorbed on the *sediments* is more difficult to characterize. In this case, there is contact with interstitial water, free water, and solid particles. A simplifying hypothesis consists of assuming that the organisms are exposed to interstitial water, so that the usual toxicity assays can be used to characterize the effects (Dickson et al., 1987). Models such as that of DiToro (1985) or more sophisticated models such as that of Boese et al. (1990), which incorporate parameters relative to the surface water, interstitial water, and sediments, take these various situations into account.

Terrestrial invertebrates

The fixation[6] of pollutants by earthworms has been modelled empirically (Wheatley and Hardman, 1968; van Gestel and Ma, 1988) or on the basis of a partition coefficient between the soil and the soil solution (Connell, 1989). The reader can find other references in Annexure 4.

Terrestrial vertebrates

Birds and other terrestrial vertebrates are exposed to polluted soil by multiple routes, whereas animals are usually exposed by oral means in toxicity assays. Rabbits, rodents, and other small mammals are exposed to polluted soils by ingestion (of contaminated soil, water, and food, such as plants, earthworms, various invertebrates), inhalation of a polluted atmosphere, or contact through the skin. The objective of the evaluator is not to obtain a detailed knowledge of the avenues of exposure at any cost, but to obtain an estimation of quantities of pollutants carried by these different means, in order to link the data relating to the occurrence and behaviour of products to data characterizing the toxic effects. In practice, and this is one of the difficulties of evaluation, that means linking the predicted or

---

[6]Fixation is the process of absorption and adsorption of the substance studied in or on the test organism (OCDE, 1981).

measured environmental concentrations in different mediums to concentrations and avenues of exposure effectively used in ecotoxicity assays. A bird may be exposed to a pollutant by inhalation or by consumption of contaminated water, plants, or insects, while in ecotoxicity assays currently used the product is administered continuously in feed or drinking water or in a single dose. It is generally impossible to estimate the parts relative to these different avenues in the total contamination of individuals, and some avenues are missed.

Daily doses for terrestrial species are estimated by relatively simple formulas of the following type:

$$DJ = [(CF \cdot FC) + (CS \cdot SC) + (CW \cdot WC)] \cdot AE/BW$$

where DJ is the daily dose in mg/kg/day, CF is the concentration of pollutant in the plants (leaves or roots) or food of animal origin, FC is the quantity of food consumed (kg/day), CS is the concentration of pollutant in the soil (mg/kg), SC is the quantity of soil ingested per day (kg/day), CW is the concentration of pollutant in water (mg/l), WC is the consumption of water (l/day), AE is the bioavailability (%), and BW is the body weight.

In the absence of direct data, the food consumption of vertebrates can be estimated according to their weight from formulas of the following type (Calder and Braun, 1983):

$$C = 59 \cdot W^{0.67} \text{ (bird)} \quad \text{or} \quad C = 99 \cdot W^{0.90} \text{ (mammals)}$$

Other such equations can be found in Tables 2.4 and 2.5. Methods for calculating exposure of birds and terrestrial mammals to fertilizers and pesticides are described by the OEPP[7] (OEPP, 1994). Some data can be transposed to cases of polluted soils and sites. Other methods for calculating exposure of birds can be found in Annexure 1 and in Gobas et al. (1995).

4.3.3. Bioconcentration and Bioaccumulation in the Food Chain

According to Ramade (1992), '*bioconcentration* is the direct growth of concentration of a pollutant as it passes from the water into an aquatic organism. This concept can be extended to terrestrial organisms in terms of passage of air and/or soils into plants,... direct passage of air into animal species by inhalation. The term *bioaccumulation* designates the sum of absorptions of a pollutant by direct and alimentary means by aquatic and terrestrial animal species.... The *concentration factor* F_c can be defined as the ratio of the concentration of a pollutant in an organism to its

---

[7]OEPP = Organisation Européenne et Méditerranéenne pour la Protection des Plante.

concentration in the biotope. In the case of PCB, for example, we have: F_c = [PCB]organism/[PCB]water or soil.... The phenomenon of transfer and biological amplification of the pollution inside contaminated biocenoses ... is called *bioamplification* or *biomagnification*.'

Here we use the term *bioconcentration factor* (BCF), presently the most widely used. There is true bioconcentration only if BCF > 1, so that the distinction between model of exposure and model of bioconcentration does not make much sense. Only the result can show whether there is bioconcentration. The distinction between bioaccumulation and biomagnification is often arbitrary. Should the contamination of a fish by consumption of aquatic plants, themselves contaminated by their contact with the water or by being rooted in the sediments, be called bioaccumulation (different sources of exposure) or biomagnification (passage across a trophic chain from water/sediment to plant to fish)?

Measurement of the bioconcentration is essential to evaluate the ecotoxicity of a product. Risk evaluations generally use BCF for fish derived from specific ecotoxicity assays (OCDE, 1981). In these tests, the final point of measurement is BCF at equilibrium in the whole fish, the edible parts (which enable us to estimate the risk for the consumer), and the non-edible parts. The rate of purification and the related residues are also evaluated. The German agency for the environment proposed the evaluation of bioconcentration by a combination of two criteria, the BCF and the rate of purification (Table 2.6; Franke et al., 1994).

The bioconcentration potential can also be estimated by indicators, the most popular of which is the octanol–water partition coefficient ($logK_{ow}$), based on a direct relation between hydrophobicity of a product and its capacity to be bioconcentrated. Most algorithms are of the following type (OCDE, 1993a):

$$logBCF \text{ (fish)} = a \cdot logK_{ow} + b$$

Other more complex equations may be used for fish, for example (Nendza, 1991):

$$logBCF = 0.99 \cdot logK_{ow} - 1.47(log[4.97 \cdot 10^{-8} \cdot K_{ow}] + 1) + 0.0135$$

The reader can find more examples, as well as algorithms applicable to other species, in the OCDE monograph (1993a).

Various researches have confirmed the existence of a relation between K_{ow} and BCF, but with sensitive differences, which can be 10 to 100 times the predicted values. According to Barron (1990), the phenomena that determine bioconcentration are the following:

- *Lipid content*: to take this into account, several authors have proposed the standardization of BCF values on the basis of lipid content in animals.

Table 2.4. Relations establishing consumption of feed (F; kg/day) and water (C; l/day) as a function of body weight (kg) and type of feed (dry or wet; according to ECAO, 1987, in Suter, 1993a; with the permission of Lewis Publishers (CRC Press), 1997)

	Equation	r	n
Dry feed	$F = 0.049 \cdot W^{0.6087}$	0.95	148
	$C = 0.093 \cdot W^{0.7584}$	0.95	148
Wet feed	$F = 0.054 \cdot W^{0.9451}$	0.97	16
	$C = 0.090 \cdot W^{1.2044}$	0.96	16

Table 2.5. Relations establishing the consumption of feed (F, g dry feed/day) as a function of body weight (W, g; according to Nagy, 1987, in Suter, 1993a)

Group	Equation	Interval of confidence (95%)[a]	n
	Mammals (uterine)		
Uterines (total)	$F = 0.235 \cdot W^{0.822}$	63–169%	46
Rodents	$F = 0.621 \cdot W^{0.564}$	64–176%	33
Herbivores	$F = 0.577 \cdot W^{0.727}$	62–161%	17
	Mammals (marsupial)		
Marsupials (total)	$F = 0.492 \cdot W^{0.673}$	37–59%	28
Herbivores	$F = 0.3211 \cdot W^{0.676}$	46–84%	12
	Birds		
Birds (total)	$F = 0.648 \cdot W^{0.651}$	55–124%	50
Passerine	$F = 0.398 \cdot W^{0.850}$	31–45%	26
Birds (desert)	$F = 1.11 \cdot W^{0.445}$	49–95%	8
Marine birds	$F = 0.495 \cdot W^{0.704}$	61–159%	15
	Lizards (iguanas)		
Herbivores	$F = 0.019 \cdot W^{0.841}$	59–146%	5
Insectivores	$F = 0.013 \cdot W^{0.773}$	30–43%	20

[a]In percentage of predicted mean value.

- *Absorption*: several physiological parameters modulate the penetration of molecules, on a basis other than a simple passive diffusion of hydrophobic molecules.
- *Steric obstacles*: several physico-chemical characteristics have an effect on bioconcentration, for example, PCBs become bioconcentrated in direct relation with their size rather than as a function of their hydrophobicity. In general, the coarse hydrophobic molecules have a lesser diffusion that does not enable us to predict their hydrophobicity.

Table 2.6. General evaluation of bioaccumulation (according to Franke et al., 1994)

	Bioconcentration factor	
BCF category	Evaluation group	Remarks
< 30	I	Low BCF
30–100	II	Average BCF
100–1000	III	High BCF
> 1000	IV	Very high BCF

	Elimination	
CT_{50} category[1]	Evaluation group	Remarks
< 3 days	I	Rapid elimination
3–10 days	II	Retarded elimination
10–30 days	III	Slow elimination
> 30 days	IV	Insignificant elimination

Overall evaluation of bioaccumulation

The various evaluation groups have the same significance in the overall evaluation:

$$\frac{\text{BCF evaluation group} + CT_{50} \text{ evaluation group}}{2}$$

If the quotient is not a whole number, we take the higher value. If the elimination data are not available, we use only the BCF groups. The final result is the classification of the product in a group and determination of an evaluation factor:

Overall group	Evaluation factor	Remarks
I	1	No risk
II	2	Potential risk
III	5	Risk
IV	10	High risk

[1]CT_{50} = clearance time.

- *Distribution*: kinetic parameters affect bioconcentration, especially in the rapidity of processes, the states of equilibrium often being reached slowly.
- *Biotransformation* can considerably reduce the bioconcentration of a product. Products such as PAH are metabolized by fish, while PCBs are only very slightly metabolized.

- *Inter-specific and intra-specific variations.*
- *Environmental conditions* such as temperature or pH for ionizable chemical species.
- *Bioavailability* in the medium, particularly difficult to estimate for highly hydrophobic molecules.

It is generally admitted that $\log K_{ow}$ values higher than or equal to 3 indicate a strong potential for bioaccumulation (for terrestrial species, cf. Garten and Trabalka, 1983). The super-lipophilic products ($\log K_{ow} > 6$, molecular weight < 600–700, or size > 0.95 nm) move with difficulty in plants and are not considered very bioaccumulative, which remains to be demonstrated more thoroughly. The ionized products pose a particular problem. The OCDE, in its directions for homologation of pesticides, recommended that the measurement of $\log K_{ow}$ be made in the non-ionized form of the molecule (that is, in conditions of maximum lipophilia).

Bioaccumulation is not always clear enough to characterize. Four variables determine bioaccumulation at each trophic level:

- concentration of pollutant in prey or the mediums;
- bioavailability;
- total quantity of pollutant carried by each avenue of exposure; and
- rate of elimination of pollutant by the predator.

Bioaccumulation can be estimated either with approximations provided by $\log K_{ow}$ or by predictive models (Barber et al., 1988; Thomann, 1989; Phillips, 1993), in addition to relative loads (measured or estimated) of different avenues of exposure.

There are very elaborate algorithms, of the type developed by Gobas for fish (Gobas et al., 1995), in which the change in concentration with time is represented by the following equation:

$$dM_F/dt = (k_1 \cdot C_{WD} + k_D \Sigma P_i C_{D,i}) V_F - (k_2 + k_E + k_M) \cdot C_F V_F$$

where M_F is the quantity of pollutant in the fish, k_1 is the aqueous flow via the gills, C_{WD} is the concentration of pollutant in water (bioavailable), k_D is the consumption, $\Sigma P_i C_{D,i}$ is the concentration of pollutant in various foods, V_F is the weight of the fish, k_2 is the rate of elimination by the gills, k_E is the rate of faecal elimination, k_M is the rate of metabolic degradation of pollutant (equal to 0 for persistent pollutants), and C_F is the concentration of pollutant in the fish.

Some of these values are themselves calculated from constants or empirical equations, for example:

$$k_D = E_D \cdot F_D / V_F$$

where E_D is $[(5.3 \cdot 10^{-8} \cdot K_{ow}) + 2.3]^{-1}$ and $F_D = 0.022 \cdot V_F^{0.85} \cdot e^{0.06 Tw}$

where Tw is water temperature.

The data, the standardized protocols, and the models of bioaccumulation apply essentially to fish, but some also exist for aquatic invertebrates (Markwell et al., 1989).

In the absence of a better alternative, BCF calculated for fish can be used to evaluate bioconcentration in terrestrial vertebrates. The data found in the literature enable us also to derive BCF for pesticides in birds and wild rodents (Mahoney, 1974; Forsyth and Peterle, 1984; Garten and Trabalka, 1983). A simple formula for risk evaluation for birds and fish-eating mammals has been developed by Romijin et al. (1993):

$$MAR = NOEC_{fish\text{-}eater} \cdot BCF^{-1}$$

where MAR is the maximum acceptable risk level (equivalent to van Straalen and Denneman's HC5; cf. Annexure 3), $NOEC_{fish\text{-}eater}$ is the NOEC value for the group of species considered (derived by extrapolation from existing data), and BCF is the bioconcentration factor in fish.

Fordham and Reagan (1991) have proposed a model for evaluating biomagnification, specially adapted to analysis of contaminated sites and to risk of bioaccumulative pollutants for species at the top of the food chain.

The published data resulting from ecotoxicity assays, microcosm assays, or measurements of concentrations in mediums and animal tissues can be used to calculate empirical factors of bioaccumulation, enabling a reasonable estimation of risk for terrestrial invertebrates and vertebrates. If the situation warrants, it is always possible to conduct a bioassay or a microcosm study with the mediums from a polluted site to have an exact idea of the bioavailability for a species at risk. For example, Shu et al. (1988) added a certain percentage of contaminated soil to the feed of laboratory rats. This method is very useful, because it allows an overall evaluation of exposure (and possibly toxic effects, if research on biomarkers is done at the same time).

4.3.4 Attraction and Repulsion

The exposure of animals can be considerably modified by attraction and repulsion. Repulsion can be considered a positive effect, but it can lead also to the abandoning of a habitat. In order for there to be an established reaction, the animals must associate the pathology and the contaminated food, distinguish between healthy and contaminated food, and live long enough to develop an aversion to the sources of contamination. In the absence of behavioural data, the scenario supposes that there is no repulsion or attraction. For additional information, the reader can consult Linder and Richmond (1990).

4.3.5. *Spatio-temporal Development of Individuals*

Although it is easy enough to obtain detailed information about the occurrence and behaviour of a pollutant at different spatial scales, we generally have less information on the density and spatial distribution of individuals of an animal population and the duration of their residence at a given place. This does not pose a problem in the case of homogeneous pollution of an environment, such as a diffuse pollution covering an area wider than that in which the animals move. In the case of a polluted site, certain species (for example, earthworms) are limited to the site, even if their movements within the site (vertical movements, for example) are not always well known. Other species (birds) frequent the site only occasionally. In this case, we generally adopt a worst-case strategy in supposing that they frequent only this site, but the exposure will obviously be overestimated.

4.3.6. *Toxicokinetic Models*

Toxicokinetic models, which aim to establish the internal dose of a toxin from environmental concentrations (external dose), have been developed most of all for aquatic species, fishes, and benthic invertebrates (Landrum et al., 1992). They have three essential functions, according to Suter (1993a):

- To estimate an internal dose when the effects are characterized on the basis of a dose-effect relation (for example, in an ecotoxicity assay, supposing the mortality is expressed as a function of the concentration of a toxin in an organ, it will be necessary to calculate these internal doses from measured environmental concentrations, or at least to have available the corresponding data by measurements on living organisms captured *in situ*).
- To integrate exposures irregular in duration.
- To extrapolate between species and stages of development.

Landrum et al. (1992) have done a complete analysis of different types of toxicokinetic models. They distinguish *equilibrium models*, currently the most often used, from *kinetic models*, which they feel are better adapted to describing exposure of animals in conditions of non-equilibrium resulting from irregular and/or multiple exposures. The kinetic models are divided into three groups:

- compartmental models
- physiological models
- bioenergy models

Compartmental models, the most simple, consider living elements as a single compartment, in which the concentrations of pollutant will be in equilibrium with the environment. Mathematically, the model is expressed usually in the following form:

$$dC_i/dt = (r_e \times C_e) - (r_s \times C_i)$$

where C_i is the internal concentration (in the organism), C_e is the external concentration (environmental), and r_e and r_s are the constants of speed of ab⁻orption and elimination. Bioconcentration models such as those of Mackay or Gobas fall within this category. With these models, we can calculate the half-life of the concentrations at equilibrium. The rate of concentrations at equilibrium is useful in the calculation of BCF, which we can use to deduce the internal concentrations from environmental concentrations. The parameters of the model are based on empirical relations or calculated by structure-activity relationships.

More complex models, comprising a storage compartment in addition to a principal compartment, have been developed for pollutants with a strong capacity for storage in body fat. A model of this type has been proposed by Kara and Hayton (1984) to simulate the kinetics of 2-ethyl-hexyl phthalate in a fish. The parameters have been calculated experimentally from laboratory assays. A model with two compartments was proposed by Clark et al. (1987, 1988) for the seagull (Fig. 2.7).

Another class of pharmacokinetic models is that of *physiological models* (PB-PK, physiologically based pharmacokinetic models), which no longer consider mathematical entities such as compartments, but real anatomical

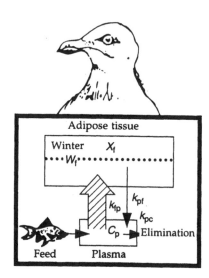

Fig. 2.7. Bi-compartmental model used to characterize the occurrence of lipophilic pollutants in the seagull (k_{pf} and k_{fp} are constants of rate of adipose tissue-plasma exchanges; X_f is concentration in the adipose tissue; C_p is concentration in the plasma; W_f is the winter level; according to Clark et al., 1988, with the permission of the American Chemical Society, 1997).

entities (such as liver or muscles) and the physiological variables responsible for transfers, such as the blood debit in the organs or rates of biotransformation. These very recent models were developed primarily for humans. One physiological model was proposed by Barron (1990) to study the occurrence of a pollutant absorbed by the gills and metabolized by the liver of fish (Fig. 2.8).

The *bioenergy models* (bioenergy-based toxicokinetic models) estimate the passage of pollutants from exchanges of matter (gas, water, food) between the external medium and the organism. These models are very similar to physiological models.

4.4. Mathematical Models

4.4.1. The Advent of Models: Advantages and Limitations

We saw earlier that the distinction between model and scenario (cf. chapter 1) corresponds essentially to the sense given to the word 'model'. Scenario

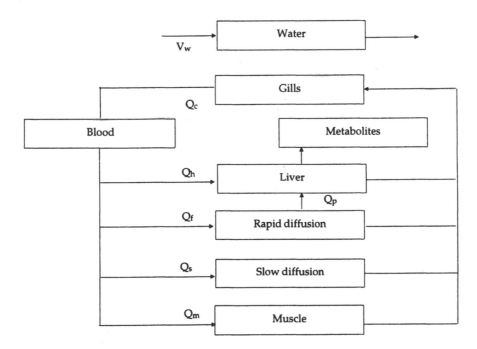

Fig. 2.8. Toxicokinetic model of occurrence of a pollutant absorbed by the gills and metabolized by the liver (V_w is volume of ventilation, Q_c is cardiac flux, Q_h is hepatic flux, Q_p is portal flux, Q_f is flow toward highly vascularized tissues, Q_s is flow towards poorly vascularized tissues, and Q_m is flow towards the muscles; according to Barron, 1990, with the permission of the American Chemical Society, 1997).

is defined as a primary model, a qualitative description of different elements in contact with each other and their relations with each other. The 'mathematical' model represents the quantitative formulation of these relations. Moreover, the complexity of mathematical models is highly variable, from simple equations (or more complex algorithms[8]) linking two or several parameters (for example, the relation between BCF in plants and the K_{ow}), which are partial models, up to more total models such as multimedia models.

Most of the existing models are predictive models of the occurrence and behaviour of pollutants. According to Calamari and Vighi (1992), this concept was introduced by Baughmann and Lassiter (1978) to define a quantitative approach that 'does not incorporate the dynamics of the particular environment, but is based on the properties of a "stylized" environment and hypothetical pollutants for which it is possible to define (rather than measure) the inputs.' Since then, there has been an explosive development of complex models that can be operated quickly by reasonably skilled users, owing to the power of computers presently available. The convenience of these models is only apparent and the best specialists have emphasized the dangers of their uncontrolled use without an understanding of the principles and hypotheses they are based on.

The principal applications of mathematical models are the following (Calamari and Vighi, 1992):
- a preliminary trial to evaluate the behaviour of numerous products;
- an indication of the affinity of pollutants for the different abiotic mediums;
- identification of compartments at which high concentrations—or, on the other hand, rapid degradation—can be predicted;
- information for the planning of sampling campaigns;
- calculation of data for a preliminary evaluation of danger.

4.4.2. Classification of Different Types of Models

The classical models of occurrence of pollutants in the environment are based on the notion of compartment. A compartment is a portion of space, large or small, for example, a section of a water course, a pond, the soil surface, within which the exchanges are more rapid than in the neighbouring compartments, so that the concentration within the compartment is considered homogeneous. Compartments are sometimes difficult to identify physically, for example, the aqueous, gaseous, and solid phases of soil

---

[8]Algorithm: 'set of formal rules expressed by a mathematical representation and corresponding to a necessary sequence; this mathematical representation' (Dictionnaire Robert).

correspond to three compartments that occupy, at a certain scale, the same volume.

The dynamics of pollutant transfer between compartments is based on an equation of mass conservation, that is, the quantity of matter (pollutant) present in the compartment is the difference between quantities entering and quantities going out. There are two types of models, equilibrium and dynamic. In an equilibrium model, the entries into the compartment equal the exits and the concentrations are calculated at equilibrium between the different phases. A dynamic model represents the development of concentrations in the different compartments, with time, from constants of speed of transfer between compartments. Recently, models based on the concept of fugacity have been developed (cf. Annexure 1).

According to Mackay and Paterson (in Suter, 1993a), the procedure described below must be followed in constructing a model of occurrence and behaviour:

- *Define the limits of the compartment.* The dimensions of the ecosystems must be taken into account. The initial hypothesis that concentrations are more homogeneous within the compartment than between compartments does not always correspond to reality. For example, the movements of a product in the soil may be slower than the passage of soil towards the water or towards the air. The compartment can be subdivided if necessary. For example, in multi-media models, an 'air' compartment, a 'water' compartment, and a 'soil' compartment can be defined, or two aerial compartments (gaseous phase and particle phase), or two aqueous compartments (a surface layer and a deep layer), or even more (considering different liquid layers, suspended particles, and sediments).
- *Define the nature and speed of different processes of formation and degradation of the pollutant.* The chief difficulty is to adapt the results of laboratory tests to the scenario, with different molecules and different conditions of the medium (physico-chemical characteristics such as pH of water or solar radiation). This adaptation can be made with extrapolation models, for example, models of structure–activity relationships.
- *Define the entries into the compartment* as a function of the source (localization, intensity, duration, frequency (and transfers possible from the source).
- *Define the speed of transfer between the different mediums.* The transfers correspond to exchange in the medium and diffusion between mediums. The rate of diffusion is calculated as a function of four parameters, the partition coefficients expressing equilibrium between the two phases, the two constants of speed that express the speed of passage of one fluid towards the other, and the surface of contact between the two compartments.

- *Resolve the system.* The simplest way is to consider the system at risk in the presence of a fixed source, with a constant emission, and to calculate the concentrations at equilibrium. A more rigorous solution to the problem consists in resolving the play of differential equations taking into account constants of speed of different processes.

Most of the existing models are *partial models* describing only the occurrence and behaviour of products in the environment. Models that are more complete quantify direct and indirect avenues of exposure: direct contact with the product from the source (polluted soil) and contaminated mediums (air and water), ingestion of contaminated plants and prey, and the possibility of bioaccumulation and metabolization of the product by primary producers and/or primary consumers. The degree of refinement and the predictive value of evaluations depend obviously on the available data and their quality.

Some models have been elaborated in the past few years to take various situations into account (cf., for example, OCDE, 1989, 1993b, c; ECETOC, 1992). The selection of a model adapted to the situation to be evaluated is difficult enough to require, in some cases, a guide to selection (Covello and Merkhofer, 1993). The difficulty of classing the models is clear in the variety of classification criteria proposed by different authors. For example, Fiksel and Skow (in Covello and Merkhofer, 1993) proposed seven criteria:
- the nature of mediums
- the geographical scale
- spatio-temporal characteristics of the source
- pollutants
- the system at risk (target populations, receptors)
- avenues of exposure
- temporal scale (present, retrospective, or prospective evaluation)

Covello and Merkhofer (1993) established the following classes of exposure models, most of them subdivided further:
- atmospheric models
- aquatic models (surface waters)
- models of the phreatic layer
- models of sloping basin
- models of food chain
- multi-media models
- exposure models
- population models

Finally, Mackay and Paterson (in Suter, 1993) defined three categories of models:
- multi-media models
- models of site rehabilitation

- models describing occurrence and behaviour in a single medium: atmospheric, aquatic, and terrestrial models, models of food chain, models of chemical accidents

The coherence of these classifications leaves much to be desired. For example, models of the food chain, classified by Mackay and Paterson as models relative to a single medium, clearly involve the transfer of pollutants between different compartments (biotic and abiotic compartments). The description of a model under one heading rather than another is often justified only by the particular importance given to the modelling of one phase of exposure in relation to others.

4.4.3. The Choice: Simple or Complex Model?

Models currently in use vary in complexity, describing one or several stages of an exposure scenario, in more or less detail. Many are hybrid constructions, juxtaposing several sub-models that have already been used. The most numerous simulate the behaviour of products in a medium (soil or water, for example) or different mediums (multi-media models) from various situations of pollution, for example, occurrence of a widely used product in defined conditions or industrial production of a certain quantity of a product. These models enable us to estimate the environmental concentrations to which living elements are exposed. Some models go further in including the biomass in a particular compartment, which allows the calculation of internal doses (the models of Mackay, for example). A more detailed description of the models available, along with examples, can be found in Annexure 1.

The advantages and disadvantages of using scenarios based on simple models have already been pointed out (chapter 1, section 2.4). The most simple models use numerical data that are easily accessible or easy to determine experimentally ($logK_{ow}$, for example). They provide quick evaluations, but their ecological realism is poor. On the other hand, more complex models, more directly linking the source and exposure without intermediate extrapolations that generate uncertainty, are better adapted to description of these particulars, but it is difficult to generalize them to other situations.

The limits of modelling occurrence and behaviour of pollutants in soils have been clearly pointed out by various authors (Roberts, 1992; I2C2, 1994). The two chief difficulties are the uncertain nature of the precipitation regimes (which can possibly be modelled by the definition of a probability function) and the heterogeneous nature of soils. Finally, the use of these models is restricted to low concentrations in the case of diffuse pollutions. Phenomena that appear at high concentrations, such as precipitation/dissolution of pollutant, modification of the microbial flora, and movements of non-miscible phases, are not taken into account.

A model is generally validated for a pollutant (cf. chapter 1, section 2.3.3), but it is only useful if it can be adapted either to other molecular structures or to other types of environment than those for which it was constructed. As emphasized by Roberts (1992), the three principal questions to be posed about the validity of a model are the following:

- Are the phenomena that control the occurrence and behaviour of the product characterized adequately?
- Does the integration of the different processes provide correct predictions about the occurrence and behaviour of the pollutants?
- Are there sufficient data on the physical and chemical properties of molecules for the model to function?

One of the chief obstacles to a more frequent use of models is the difficulty of obtaining reliable data (Mill and Walton, 1987). Suntio et al.'s (1988) compilation of values of Henry constants for pesticides shows that published data are often imprecise. Theoretically, only some constants are essential to the functioning of a fugacity model, such as the partial vapour pressure of the molecule and its solubility in water. Other useful physical properties can be deduced from these constants, but at the cost of extrapolations that add to the uncertainty, or propagation of a systematic error. Another constraint is the time needed for measuring the parameters of models. The EXAMS model takes into account 14 parameters relative to the pollutant and to the mediums: ionization, Henry constant, vapour pressure, aqueous solubility, sediment–water partition coefficient, organic carbon content, standard sediment–water partition coefficient, octanol–water partition coefficient, rate of oxidation, rate of photolysis in aqueous medium and quantum yield of photolysis reaction, molar absorptivity, rate of hydrolysis, and rate of biotransformation. Donaldson (1992) estimated that it would take 10,000 years at the EPA laboratories, using the means currently available, to calculate these constants for the 10^7 existing molecules inventoried by *Chemical Abstracts*.

Whatever the apparent complexity of models, it must be emphasized that they are only simulations, working with a considerable number of approximative parameters, with a sometimes redundant use of the same parameter, such as $\log K_{ow}$. The results obtained depend on the model and may be very different. As seen in the example in Figure 2.9 (Asher et al., 1985), the concentrations of hypothetical compounds at equilibrium are very different, depending on whether the model used is the fugacity model of Mackay or the persistence model of Roberts. Divergences also appear in the time needed to reach equilibrium. In the same mediums (sediments, fish, and suspended particles), the equilibrium concentrations 2,3,7,8-tetrachlorodibenzo(p)dioxin are reached in a few hours according to the Mackay model, and in years according to the Roberts model (Roberts, 1992).

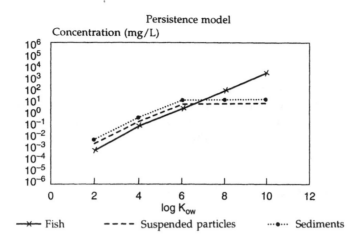

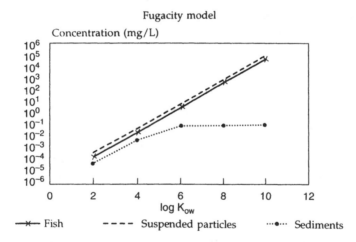

Fig. 2.9. Estimation of concentrations of hypothetical pollutants in sediments, suspended particles, and fish, as a function of their $logK_{ow}$ (according to Asher, 1985; with the permission of SCOPE, 1997)

Travis and Hattemeyer-Frey (1988) have made a remarkable demonstration of the approximative nature of models. These authors compared the fixation of trichloroethylene by plants through two different avenues, by aerial means and by the root, based on concentrations in the soil and the air. The BCF (soil–plant) was estimated by the formula:

$$B_v = 38.9 \cdot K_{ow}^{-0.58}$$

and the BCF (air–plant) by one of two formulas:

(a) $9.1 \cdot K_{ow}^{0.71}$ or

(b) $5.0 \cdot 10^{-8} K_{ow}^{1.5} / H$

The calculations showed that if we use formula (a), we conclude that fixation by aerial means represents more than 99% of the total fixation, whereas if we use formula (b), we conclude just the opposite, that aerial fixation represents less than 2% of the total fixation. This example is even more remarkable in that the value of the regression coefficient for the five model molecules used to define the model is the same in the two cases ($r = 0.97$).

4.5. *In situ* Indicators of Exposure

The exposure of living organisms is directly accessible to the evaluator by measurement of pollutant concentration or of modification of various biochemical and cellular parameters (response of biomarkers of exposure) in the tissues of the contaminated organism. For the evaluator, these measurements have several functions:
- to characterize the degree of soil pollution, particularly the fraction bioavailable to plants and animals;
- to characterize the risk to humans or predator animals of consuming contaminated plants or animal products; and
- to characterize certain long-term risks for organisms in which these measurements are made.

The choice of these options depends on the objective of the evaluation. The last point specifically concerns the characterization of effects.

The advantage of measuring concentrations of pollutants in living elements, whether plants or animals, is that it provides a direct evaluation of degree of food contamination or exposure of the species at risk, thus avoiding the uncertainties associated with indirect methods. However, we have seen that this measurement was not always necessary, since toxicity tests are based exclusively on relations between environmental concentrations and toxic effects. Moreover, the populations that are at greatest risk are often those that are low or declining in numbers (the extreme case is that of species on the way to extinction). These species benefit sometimes from a statute of protection and it is for them that samplings in nature are most difficult to take. A useful solution consists of taking substitute species, sentinel animals or plants, as final points of measurement.

4.5.1. *A System of Sentinel Organisms*

A system of sentinel organisms is defined as 'a mechanism devised to collect, systematically and regularly, data on animals exposed to environmental pollution. These data are then analysed to identify the potential dangers to human and environmental health' (National Research Council, 1991). Here, we use the term *sentinel organism* for any living element collected on a site, even if the collection is not strictly 'systematic and regular'.

During the 19th century, when miners returned to a mine after a firedamp explosion, they carried with them a canary. Signs of asphyxiation in the bird, twenty times as sensitive as a human to carbon monoxide, warned them of the presence of dangerous concentrations of this gas in the shafts.[9] The canary is no longer used, but other systems of sentinel animals are. Mussel Watch, one of the earliest, was set up in the United States around 1975 to evaluate the quality of sea water. Concentrations of pollutants are very low in sea water and therefore difficult to measure precisely. The alternative is to analyse pollutants in aquatic organisms, where they bioaccumulate sometimes to high concentrations. Another advantage over measuring levels in water is that the organism's processes of bioaccumulation and purification take some time, which eliminates momentary variations and allows a long-term follow-up (the absence of a quick response can also be considered a disadvantage: some systems of sentinel organisms, as in the case of the canary, are developed to detect the occurrence of a pollution quickly). In the Mussel Watch system, the molluscs were collected systematically during the winter from around 200 sites, then stored below $-110°C$. The size and weight were measured, as are the concentrations of several persistent pollutants (heavy metals, organochlorine molecules), and more recently various biochemical parameters. In France, the Reseau National d'Observation (RNO) of the IFREMER is based on 110 stations, in which mussels are collected in February, May, August, and September. The advantages of sentinel species, domesticated or wild, have been highlighted on several occasions (Buck, 1979; National Research Council, 1991), but there are no mechanisms equivalent to Mussel Watch for terrestrial ecosystems.

4.5.2. Which Species to Collect and under What Conditions?

The study of *vegetation* (spontaneous or cultivated on the site) is an important source of information on the bioavailability of pollutants and their transfer through the food chain.

Sentinel animals can be found in several zoological groups. The *terrestrial invertebrates* (earthworms, insects) and *aquatic invertebrates* (shellfish, crustaceans) are generally abundant and easy to collect. *Reptiles, amphibians,* and *birds* are less abundant, more difficult to collect, and most of the time considered species at risk, victims of adverse effects of environmental

---

[9] In this historical example, there is no measurement of environmental concentrations of CO_2 in the animal, but a threshold of environmental concentration is evaluated by the appearance of a toxic effect.

pollution. Certain bird species, such as seagulls or pigeons, can serve as sentinel organisms. Fish and domestic animals (ovines and cattle) are good sources of information on the degree of pollution of mediums that they frequent. Small mammals are generally abundant, with limited movement. They are closer to humans in the evolutionary scale than invertebrates or birds and are good candidates for the monitoring of terrestrial ecosystems (MacBee and Bickham, 1990; Talmage and Walton, 1991). Several animal species can be used for the study of atmospheric pollution (Newman and Schreiber, 1984).

A 'good' sentinel animal species must satisfy several conditions (O'Brien et al., 1993):
- relatively easy capture;
- sufficient population density;
- known area of distribution;
- sufficient size;
- known avenues of exposure;
- possibility of devising control animals;
- existence of other studies on the same species.

4.5.3. *Mechanisms of Exposure to Pollutants*

In a system of sentinel organisms conceived according to a *direct* approach, the data are obtained on a population living on the site. For animal species, the variants revolve around the treatment of individuals collected, which may be sacrificed immediately, brought to the laboratory for a short period, or even released immediately after the sampling of tissues or measurements of morphological, biochemical, or other parameters. In an *indirect* approach, the organisms, individuals of natural origin or cultivated from laboratory colonies, are transported to the site for a short or long period.

Systems of enclosure correspond to a well-defined mechanism of aquatic ecotoxicology. A portion of the natural space is delimited, autochthonous or exotic species are introduced in it, and the biological effects are observed after a period of their residence. These mechanisms apply to terrestrial plants (test cultures on a polluted soil), aquatic plants, underground terrestrial species (earthworms buried in the soil inside a perforated box), or caged and sometimes even uncaged birds on a site. A very interesting approach consists of exposing under these semi-natural conditions individuals of the same species used in standard assays.

An *impact* study looks at the effects of an experimental treatment on local fauna. This type of experimentation is valuable for a prospective evaluation, but does not appear to be so for a retrospective evaluation.

4.5.4. *What is to be Measured (or observed) in the Organisms Collected?*

Two types of information are looked for from sentinel organisms (or species at risk): the presence of pollutants in the tissues and the response of biomarkers of exposure.

In the French RNO and Mussel Watch, the presence of heavy metals, PAH, and organochlorine compounds is measured. A systematic follow-up of pollutant levels analogous to Mussel Watch has very rarely been done for terrestrial species. One example was the measurement of capacities of bioconcentration of zinc in different species of small mammals (shrews, voles, field mice), which were then classified according to their capacity to detect metallic contamination in soil (Talmage and Walton, 1991).

A final category of indicators is made up of *molecular and cellular alterations* caused by the pollutant, which ultimately lead to toxic effects. These *biomarkers* are studied in detail in the following chapter.

Characterization of Effects

1. DIFFERENCES BETWEEN HEALTH RISK EVALUATION AND ECOLOGICAL RISK EVALUATION

Let us recall that ecological risk evaluation differs from health risk evaluation in two major ways:

- the human species is the sole object of health risk evaluation, while ecological risk evaluation concerns the millions of species that inhabit the biosphere.
- ecological risk evaluation concerns all levels of organization of living systems, with a particular interest in effects manifested in communities and ecosystems.

However, the individual remains the focus of ecological risk evaluation (cf. chapter 1). The role of the individual is clearly affirmed by the structure of ecological risk evaluations, which are almost always based on the results of monospecific ecotoxicity assays. The essence of the effort, in the logic of these tests, consists of developing methods that allow *extrapolation* to the very different conditions of the natural environment: first to the millions of species actually present in the biosphere and then to communities and ecosystems, that is, to relations of these populations among each other and with their environment. It is a characteristic difference from health risk evaluation, in which interspecific extrapolation is done from some laboratory species to a single species, man.

As for exposure, different approaches are possible for evaluating the biological effects of stresses, corresponding to the three already described, testing, modelling, and monitoring.

Here we distinguish:

- *Experimental models*, from the simplest to the most elaborate. We will use the terminology of Leon and van Gestel (1994), who distinguished between laboratory ecotoxicity assays, in which living elements are exposed to pure products, and laboratory bioassays, in which living elements are exposed to mediums taken from the site (soil or water), either directly to the contaminated mediums themselves (daphnia or fish are submerged in water taken from the polluted site or dilutions of this water), or to aqueous or organic extracts of these mediums, added to culture mediums (water for daphnia or fish) or to the feed. The protocols of standardized bioassays are modelled, in general, on those of the ecotoxicity assays from which they are derived. The *integrated tests* (microcosms, mesocosms, etc.) constitute another category of experimental models, more elaborate but less standardized.

- *Mathematical models* of different types, models of test results, extrapolation models, and simulation models.
- In situ *indicators*, corresponding to an eco-epidemiological approach that is either *direct* (plants or animals collected on the contaminated site) or *indirect* (animals caged or plants cultivated on the site).

2. LABORATORY ECOTOXICITY ASSAYS (MONOSPECIFIC TESTS) AND BIOASSAYS

The ecotoxicity assays conventionally used to measure the toxicity of chemical products are *monospecific ecotoxicity assays*, which establish a relation between a dose of substance administered under defined conditions and the response of an individual plant in a population or a given animal. The term ecotoxicity *assay* has been used in the rest of this section, but the ideas developed apply also to *bioassays*.

These assays are mentioned here briefly. The reader can find details about conducting them in several works and reports such as the *Handbook of Ecotoxicology* (edited by Calow, 1993, 1994) and *Ecological Assessment of Hazardous Waste Sites* (Warren-Hicks et al., 1989). Most of them have been standardized by directives or in the course of research, thanks to the efforts of various national and international organizations, such as the OCDE, EPA, ISO, AFNOR,[1] or recognized scientific organizations, such as SETAC.[2] It is recommended that the reader consult these documents, for example AFNOR (1995). The assays, in most cases, follow good laboratory practices (GLP: cf. OCDE, 1992a). For various reasons, essentially the delayed interest in terrestrial environments and the difficulty of the task, standardization of aquatic ecotoxicity assays is much more advanced than that of terrestrial ecotoxicity assays. Warren-Hicks et al. (1989) distinguished assays of class I from those of class II. The class I assays represent

---

[1]ISO, International Standardization Organization; AFNOR, Association Française de Normalisation.
[2]SETAC, Society of Environmental Toxicology and Chemistry.

standardized methods, defined by directives, while class II assays are those for which there is no precise protocol (for example, the arrangements for caged animals). This distinction is not a value judgement—both types of tests are useful in risk evaluation—and has not been explored here.

Evaluation is generally based on a set of assays involving species belonging to different trophic levels and inhabiting different biotopes.

The animal and plant species used for these tests are selected on the following basis:

- their role in biocenoses is significant;
- their biology is well known;
- they are not very susceptible to genetic variations (their sensitivity to toxins is sufficiently invariable so that reproducible results can be obtained);
- they are easy to maintain and available throughout the year;
- they can be associated within model ecosystems or food chains.

All these criteria are far from being fulfilled in practice and many species are used essentially for reasons of convenience (cost, availability, maintenance) rather than for reasons of ecological realism. The Japanese quail used in ecotoxicity tests on birds has no equivalent in the wild (unlike the mallard duck), but the fact that it can be raised in large numbers ensures the availability of relatively identical animals throughout the year. Among fish, the trout is an example of a noble species, of great value, and representative of an unpolluted aquatic environment. For this species, laboratory ecotoxicity assays have an ecological significance.

The criteria listed above are not all equally important. In order to ensure the reproducibility of results among laboratories, the most important characteristic is the homogeneity of the animals used. This emphasis is justified, for example, when we want to prove very slight differences in toxicity between two chemical products, but it can be questioned when results are extrapolated to natural conditions, where variability is the rule. Moreover, variability remains high even in the best-calibrated assays (Forbes and Forbes, 1994).

The species commonly used belong to large groups of the animal and plant kingdoms:

- bacteria
- algae
- terrestrial and aquatic plants
- terrestrial and aquatic invertebrates
- fish
- birds

The rat and other laboratory mammals (rabbit, for example) are also good models for evaluating risk for terrestrial mammals.

The principal final points of measurement are *mortality* (for all animal species), *growth inhibition* (algae and plant species), and *reproduction* (daphnia, fish, birds). Bioaccumulation can be measured for aquatic species. These assays correspond generally to three durations of exposure:

- assays of short-term toxicity (acute toxicity);
- assays of medium-term toxicity (subacute toxicity, duration of exposure at least 10% of the animal's lifetime);
- assays of long-term toxicity (chronic toxicity, duration of exposure equal to one reproductive cycle of the species studied).

To study long-term exposures, three methods are available:

- *static assays*: in which the medium is not renewed;
- *semi-static assays*: in which the medium is renewed at regular intervals;
- *dynamic assays*: in which the medium is constantly renewed.

The advantages and disadvantages of these methods depend on the stability of the substance and the results desired from the test. A static method is easier to conduct than a dynamic method (less substance is used, less maintenance is required) and, in those conditions, the substance degrades 'naturally', which can be considered an advantage or a disadvantage.

Short-term exposures serve to characterize only mortality, while assays of long-term toxicity allow the testing of effects on growth and reproduction as well.

As opposed to *conventional assays* done with standardized protocols, *parametric tests* indicate the consequences of variations in different parameters, the value of which has been fixed precisely in standardized tests (cf. chapter 1, section 2.3.1). In conventional tests on fish, for example, it is useful to fix the temperature at a single value (or two values at most, depending on whether the species are cold-water or warm-water species) and then to apply factors of correction to adjust the values of toxicity according to actual ambient temperatures. The parametric test enables us to establish the temperature–response relation and thus calculate a correction factor that is a little more precise in the scenario if the evaluator finds that useful (cf. chapter 3, section 4.3.9).

Bioassays are laboratory assays using the polluted mediums themselves, or dilutions or extracts (aqueous and organic) of these mediums, to expose plants or animals. They are particularly useful, in terms of ecological realism, to complete eco-epidemiological data and chemical analyses in an evaluation of polluted sites. The use of single tests is insufficient. It is recommended that a series of tests be done, designed to represent the diversity of plant and animal communities, especially the possible differences in sensitivity of species to toxins (Greene et al., 1988b; Giesy and Hoke, 1989; Munawar et al., 1989; Weber et al., 1989). We have seen earlier that the conceptual basis of bioassays is essentially the same as that of standardized assays. The most important differences that make a bioassay more complicated than a classic assay lie in the definition of principles and criteria of preparation of mediums (polluted water or soil) to which the living organism is exposed.

Several series of bioassays have been proposed (Reynoldson and Day, 1993; Giesy and Hoke, 1989; Greene et al., 1988b; Weber et al., 1989). The Canadian approach to selection of bioassays to be used in an ecological risk evaluation of polluted sites is presented in greater detail in Annexure 2. Other batteries of tests have been proposed (e.g., Thomas et al., 1986; Athey et al., 1989; DECHEMA, 1995), but it is not within the scope of this work to suggest a particular choice among them.

2.1. Aquatic Organisms

2.1.1. Bacterial Assays

Bacterial assays have several advantages: they are economical, can be done quickly, are simple to use, and are easy to miniaturize and automate (Block et al., 1989; Reteuna et al., 1989; Mayfield, 1993). Various final points can be measured:

- variation of bacterial growth;
- measurement of enzymatic activity (Toxi-Chromotest^{TM});
- state of energy reserves;
- inhibition of luminescence (Microtox^{TM});
- consumption of substrate (organic molecule, oxygen); and
- formation of a terminal metabolite (CO_2, methane, nitrites).

Among the most often used standardized bacterial tests, and the most useful, is the Microtox^{TM} test, based on the inhibition of luminescence in a marine bacterium, *Photobacterium phosphoreum* (Vasseur et al., 1984; Lambolez et al., 1994). The more recent Toxi-Chromotest^{TM} uses a mutant of *Escherichia coli*: the toxins cross the bacterial wall and block the synthesis of b-galactosidase, the activity of which is indicated by a colorimetric method.

The chief criticism against standardized bacterial tests is their lack of ecological realism. They can be considered useful for characterizing exposure (presence and possibly measurement of quantities of toxic products in a medium) rather than for predicting toxic effects.[3]

---

[3]The test of alteration of a bacterial genome (Ames test) is a typical example of a test designed to indicate toxic properties of substances but having no ecological significance, because there is no ecological receptor identified and because, as a general rule, the possibility of mutagenic or carcinogenic effect is presently considered secondary in ecological risk. It is for this reason that the test is used less often in ecological risk evaluation than in health risk evaluation.

2.1.2. Assays on Algae and Vascular Plants

Algae can be considered for use in chronic toxicity assays, because the growth of algae in liquid nutritive mediums is rapid (most tests take 3–5 days). The final point of measurement is generally growth inhibition (Lewis, 1993). Several species are used, such as *Selenastrum capricornutum*, *Scenedesmus subspicatus* (the most often used), *Chlorella vulgaris*, and *Microcystis aeruginosa*.

Toxicity assays on algae have been adapted specifically to effluents (Greene et al., 1988a, b) and to polluted soils (Ireland et al., 1991).

The duckweed (*Lemna minor* and *L. gibba*) and many other species such as *Elodea canadensis* and *Vallisneria americana* are used in evaluation of aquatic toxicity in vascular plants. Exhaustive information on the various phytotoxicity assays currently used, their advantages, and their disadvantages can be found in a review by Wang and Freemark (1995).

2.1.3. Assays on Aquatic Invertebrates

The best known are the assays done on different species of small aquatic crustaceans, such as daphnia, but other species also can be used (Persoone and Janssen, 1993; see, e.g., Annexure 2). An abundant literature has been devoted to the standardization of daphnia tests, particularly by the OCDE. Two types of tests are currently used, one for acute toxicity and the other for chronic toxicity.

In acute toxicity assays, generally on *Daphnia magna* or *D. pulex*, the animals are exposed to a polluted aqueous medium for 24 to 48 h. At the end of this period, the immobile animals are counted and the results expressed in the form of LC_{50} or NOEC according to the standard methods.

In chronic toxicity assays developed with *Ceriodaphnia reticulata* or *C. dubia*, the animals are exposed to polluted medium for a period of 7 days and the rate of survival and reproduction is calculated. With *D. magna* or *D. pulex*, the animals are exposed for 14 to 21 days and the rate of reproduction is calculated. The results are expressed as number of descendants in relation to the control batches.

Many other invertebrate species can be used, but their growth conditions and their sensitivities to different toxins are obviously less well known.

2.1.4. Assays on Vertebrates

The assays on vertebrates essentially involve fish (Solbe, 1993). The principal species are rainbow trout (*Oncorhynchus mykiss*), sunperch (*Lepomis macrochirus*), zebra fish (*Brachydanio rerio*), or fathead minnow (*Pimephales promelas*). The tests are for short-, medium-, and long-term toxicity, in static, semi-static, or dynamic conditions. In fish, particular directives have been elaborated to test the possibilities of bioaccumulation or the effects on juveniles.

FETAX[TM] is a particular test that uses the response of an amphibian (*Xenopus laevis*) to detect mutagenic products. Its objective is not to evaluate the risk of pollutants for amphibians, but to detect a particular kind of risk (mutagenicity; cf. section 2.1.1).

2.2. Terrestrial Organisms

Ecotoxicity assays applicable to terrestrial organisms are much less developed than aquatic ecotoxicity assays. To fill in the lacunae, several research programmes have been launched in European countries, notably the Netherlands (NISRP, Netherlands Integrated Soil Research Program; Eijsackers, 1989), Sweden (MATS, Mark Test System; Rundgren et al., 1989; Bengtsson and Torstensson, 1988), and the European Community (SECOFASE, Sublethal Effects of Chemicals on Fauna Soil Ecosystem; Lokke and Van Gestel, 1993). A very interesting review on the methods that can be used to evaluate the effects of pollutants on soils has been published by Verhoef and van Gestel (1995).

2.2.1. Bacteria and Soil Microflora

Bacteria and soil microflora are often considered not autonomous living organisms, but rather integral constituents of a living system, the soil. All else being equal, the soil bacteria can be assimilated to the intracellular organelles, such as mitochondria, which ensure well-differentiated functions, but which are difficult to observe outside the cellular structure they are part of. The tests for biological activity of soil depend on structural or functional characteristics measured in the laboratory (van Straalen and van Gestel, 1993). Structural parameters such as the following can be measured:
- abundance of microflora and microfauna (biomass), by direct counting or various biochemical techniques;
- taxonomic wealth of microflora and microfauna.

Among the functional parameters that can be used are the following:
- respiration and mineralization of organic substrates;
- ammonification and nitrification;
- nitrogen fixation;
- denitrification;
- enzymatic activities: e.g., urease, dehydrogenase, phosphatase.

Compared to aquatic ecotoxicity assays, these assays are more difficult to standardize because they require a typical soil, colonized by a standard flora and fauna, which is a problem technically (a standard soil is much more difficult to obtain than a standard liquid medium) as well as theoretically (it would be necessary to define in what way one soil is more representative than another).

2.2.2. Soil Micro- and Macrofauna

Soil fauna is extremely rich in protozoa and invertebrates. The principal groups represented are the protozoa, nematodes, isopods, diplopods, and acarids (oribatids). Among the epigaeal insects, the most often studied are the collembola (e.g., *Folsomia candida*). In the macrofauna, we can cite the enchytreid, the earthworms, and the molluscs. The species most often used for ecotoxicity assays are earthworms (*Eisenia foetida, Lumbricus terrestris*) and the collembola (Bouche, 1988; van Straalen and van Gestel, 1993; Jepson, 1993; Verhoef and van Gestel, 1995).

2.2.3. Terrestrial Plants

Various final points are used to evaluate the phytotoxicity of products. The two principal ones are seed germination and root elongation (Kaputska and Reporter, 1993; OCDE, 1984). These tests are done with different vascular plants: e.g., lettuce, cabbage, carrot, wheat. Assays on the complete biological cycle of *Arabidopsis* sp. and *Brassica* sp. have been proposed. Other parameters can be measured, for example, physiological variables such as the rate of photosynthesis (fluorometric analysis of rates of chlorophyll), the rate of gaseous exchanges, and nitrogen fixation capacity. Genotoxicity assays have been developed on different plants (e.g., *Tradescantia* sp.).

2.2.4. Vertebrates: Birds and Mammals

There are ecotoxicity assays for birds, based on some species that are easy to propagate, such as the Japanese quail (*Coturnix coturnix*) in Europe, the Northern bobwhite (*Colinus virginianus*) in America, or the mallard duck (*Anas platyrhynchos*). The assays normally used are those for acute toxicity by oral route (over 24 h), for sub-chronic toxicity by alimentary route (over 5 days), and effects on reproduction (Walker, 1993; Smith, 1993), according to OCDE norms. These assays have been developed essentially to test the effects of pesticides.

There are no assays specifically done for terrestrial mammals, but the data acquired with the classic laboratory rodents and rabbits are valuable sources of information.

2.3. *In vitro* Tests

Here we define *in vitro* tests as assays on living elements at levels of organization lower than the organism, for example, purified enzymes or plant or animal cell cultures. Assay kits are available in the market to test for the presence of organophosphorous or carbamate insecticides in water. They are based on the inhibition of cholinesterase, a purified animal enzyme. The significance of these tests is ambiguous. Their primary objective is to measure the concentration of a product and, logically, they must be classified with the methods of chemical analysis of pollutants. However, cholinesterase is a molecular structure arising from a living element, and its inhibition can be interpreted as the first manifestation of a toxic effect.

2.4. Toxicodynamic Models

Toxicodynamic models are defined here as a particular class of ecotoxicity assay aimed to link the *internal dose* to a given toxic effect in the individual. We find in this category toxic effects from concentrations (measured or estimated) of the pollutant in the organism. These models can be useful to complement models of occurrence and behaviour of products that provide estimations of pollutant concentrations in the biomass, such as the Mackay model. Another category of models is based on the *mechanisms of action* of a toxin and incorporates a toxicokinetic compound describing the occurrence of a pollutant up to the molecular target responsible for a mode of toxic action. The conception and use of these models presupposes a detailed knowledge of the mode of action of pollutants. The development of the models is particularly complex; it is an avenue of research in ecotoxicology that very clearly goes beyond the present objectives and possibilities of ecological risk evaluation.

3. MODELS OF RESULTS

Conventionally, a toxicity assay aims to establish a dose–effect relation. It progresses according to the following scenario:
- a group of animals or plants is divided into several lots;
- each lot is exposed to a different dose of the toxin in defined conditions; and
- a biological effect is observed at the end of a period fixed beforehand.

Even the most simple ecotoxicity assays are based on complicated scenarios. The complete description of an experimental protocol often runs several pages and many factors can modify the result. The final representation of the relation between exposure and effect must incorporate the various dimensions needed for the subsequent evaluation: dose (quantity of product, duration, frequency), magnitude of the effect, and proportion of individuals affected.[4] The complete description of results is too complex

---

[4]The models used in toxicology to represent the dose–response relation are based on a binary relation (response or absence of response, for example, death of an individual or no death). They cannot be adapted to different situations, for example, when the response is continuous (measurement of an enzymatic activity or a rate of growth).

to be used directly and most of the time the information must be summarized as a single value representing the set of results. This single value is obtained by applying different mathematical models, based on hypotheses of a biological nature.

3.1. Dose–Effect Relations

The model most often used to represent the dose–effect relation is the classic log(dose)–integer function:

$$P = a + b \cdot LogD$$

where P is percentage of responses in the sample (transformed into integer), D is the dose, and *a* and *b* are constants. The function supposes that the distribution of individual responses has a uniform quantity of toxin following a statistic log-normal law, which is far from having been proved. Other models are used, but less frequently, such as the logit function or the Weibull function.

3.2. Final Points of Measurement

The final point of measurement most often used in ecotoxic risk evaluation is LC_{50} or LD_{50}, generally calculated from the log(dose)–integer relation. To quantify the uncertainty, the LD_{50} must be accompanied by the standard difference of its value. The LD_{50} is not always the most useful value, because it is often necessary to calculate the doses responsible for much weaker effects, such as LD_{10} or LD_{01}. It is possible if one knows the vertical slope representing the log(dose)–integer relation but, unfortunately, this information is not always available in the literature, especially in the older studies.

It is not possible, for theoretical reasons, to estimate a dose without effect (an LD_0) from log(dose)–integer relations. Such determinations are made directly from raw results (data of mortality for each lot).

The LOEC is the lowest concentration tested for which a significant effect is observed, with respect to control, and NOEC is the concentration tested immediately lower to LOEC.[5] Contrary to LD_{50}, which is a *calculated* value, NOEC and LOEC are *experimental* values, depending especially on intervals between the doses effectively tested and the size of samples. With a small number of doses, we can easily define a NOEC and a LOEC, but not very precisely. If the *number of samples* is increased to have closer intervals, we reduce the interval between the LOEC and NOEC. By that means we increase the precision of the measurement but do not avoid

---

[5]The notions of NOAEL (no observed adverse effect level) and LOAEL (lowest observed adverse effect level) are not used in ecological risk.

variability and indeed we risk obtaining, in a certain range of doses, a 'random' distribution of lots in which there may be either zero deaths or one death, and the precision is finally not improved. If the *size of samples* is increased, the interval between the NOEC and LOEC is not modified, and the precision will be the same. However, the probability of observing *one* individual presenting signs of toxicity increases in each lot and the NOEC and LOEC values will tend to decline towards the weakest values.

The MATC (maximum acceptable toxicant concentration) is defined as the geometric mean of the NOEC and the LOEC for a long-term toxicity assay (reproduction, growth, etc.; Rand and Petrocelli, 1985). This value is often considered a threshold of toxic effect (Suter, 1993a).

The preceding definitions are the ones generally accepted, but sometimes the values of final points are derived from each other by arbitrary relations, for example, the LOEC will be equal to LC_x (16 < x < 25), the NOEC will be equal to LC_x (x < 16), LOEC/2 or $1/\sqrt{2}$ MATC (ECETOC, 1993).

The other types of relations, for example, between duration of exposure and response (for a fixed dose), are less frequently represented.

4. EXTRAPOLATION MODELS

4.1. Why Extrapolate?

Suter (1993a) remarks that it is not possible to test experimentally all the toxic effects of all the products in all the possible conditions and on all the animal and plant species that exist. As we have seen earlier for exposure, risk evaluation starts with models and existing data (final points of measurement) and attempts to simulate as well as possible the present complexity and future evolution of natural environments, by extrapolating this information to real ecological values (final points of evaluation). These practices are not unanimously accepted, because extrapolation adds to the uncertainty, but they are inevitable. The uncertainty can be quantified by different methods. The most remarkable methods, based on the analysis of probability density functions of different parameters, were developed by the research team of Suter under the term 'analysis of extrapolation error' (AEE; Suter et al., 1983, 1986; Suter, 1993a; Logan and Wilson, 1995). They are particularly useful in quantifying the uncertainty associated with risk, but their complexity and above all the number of data required go beyond present needs.

Peakall and Tucker (1985) remark that 'the quantity of extrapolation one decides to incorporate in the scenario is in direct relation to the level of uncertainty one is willing to accept.'

4.2. Different Methods of Extrapolation

Two groups of methods are used to extrapolate the results of standardized toxicity assays to natural (real) situations.

The first comprises simple mathematical models of the additive or multiplicative type. These arbitrary correction factors are designated as safety factors, factors of application, or factors of evaluation. A factor of evaluation has been defined as 'the expression of degree of uncertainty accompanying the extrapolation of results of assays done on a limited number of species to the real environment. The more numerous the data and the longer the assays, the more the degree of uncertainty and factor of evaluation are reduced' (Rule 1488/94 of the CEE, 1994).

Several factors can be applied simultaneously or successively, by addition or by multiplication. A classic example of this procedure is the series of correction factors that can be used to evaluate the health risk of chemical products as a function of the nature and quality of available data (Table 3.1). Each factor 10 represents an extrapolation and the final safety factor is obtained by multiplication of several factors 10. These correction factors apply to characterization of effects (for example, to be extrapolated from one species to another) as well as risk characterization. The scientific justification of these simple models is often very weak. At best, they are expert judgements.

Table 3.1. Safety factors that can be used for health risk evaluation

---

- factor 10 to extrapolate from the average individual to the human population as a whole
- factor 10 to extrapolate from animal to human
- factor 10 to extrapolate from acute NOEL to chronic NOEL
- factor 20 to go from LOEL to NOEL
- additional modifying factors, with values from 1 to 10, take into account other uncertainties (incomplete data, limited number of species tested)

The second way to derive correction factors is to base them on statistical models (especially regression models), which give more precise values. For example, if we have data showing that a species at risk is around 12 times as sensitive to the entire range of organic products as the species used in the ecotoxicity assays, we apply a correction factor of 12 to forecast the toxicity for the species at risk from a new product.

In every case, we can systematically use a worst-case strategy (cf. chapter 1, section 2.4), that is, a conservative approach for each parameter to be estimated. For example, to evaluate the risk of a toxic effluent on fish, the most sensitive development stage of the most sensitive species will be exposed to the most concentrated effluent, in conditions of weakest foreseeable water flow. This situation has the advantage of ensuring the minimum risk level, but it is controversial because it is excessively cautious, whereas risk evaluation proposes to be not only reasoned but also reasonable.

Extrapolation is designed to estimate the importance of variables such as age, size, and development stage in the final expression of toxicity, but it is not always necessary to extrapolate all the parameters of the model. Once the significance of the variability introduced by a parameter in the result is estimated, we can decide whether to incorporate it in the final diagram. We can estimate, for example, that variations due to the exposure of animals of different ages are negligible compared to variations due to the exposure of animals of different species and, on the other hand, we can take into account the age parameter if it is established that the product has marked effects on juveniles.

4.3. Principal Extrapolations from Monospecific Assays

4.3.1. Dose

In many natural situations, we must evaluate the impact of exposure to doses weaker than the $LD(C)_{50}$. This does not pose a problem if one knows the vertical slope linking the log-dose to the integer. The variation of the result will generally be increased and we calculate it from the variance of the LD_{50}. Unfortunately, these data are not always available and in many cases we must estimate the value of the slope from an analysis of published data. For example, the Office of Pesticide Programs of the EPA has estimated that, in most tests, a toxic dose equal to one-fifth of the $LD(C)_{50}$ corresponds to $LD(C)_{0.1}$ and in extreme cases to $LD(C)_{10}$ (Urban and Cook, 1986).

4.3.2. Duration of Exposure

Acute toxicity assays are established in short durations fixed according to limits, for example 24 or 96 h. This duration allows the assay to be completed without feeding or reproduction of the organisms. The duration of chronic toxicity assays is variable and is generally calculated so that the toxic effect measured has time to manifest itself.

Under natural conditions, exposure can be shorter or longer, sometimes irregular. The irregularity of exposure is due either to the source of pollution or to the population exposed. There are four possibilities, in order of growing complexity:
- a regular emission and a continuously exposed population;
- an irregular emission and a continuously exposed population;
- a regular emission and a discontinuously exposed population;
- an irregular emission and a discontinuously exposed population;

Various methods have been used to characterize the importance of this factor. Bailey et al. (1984) developed equations to convert the LC_{50}–96 h to LC_{50} for exposures of 1–8 h and 24–72 h. Another possibility is to base it on a concentration-duration relation. The most simple model postulates that the product of the toxin concentration and the duration of exposure is a constant:

(1) $c \times t = k$

where c is the concentration of product, t is the duration of exposure, and k is a constant.

In other words, for the same concentration, a product is twice as toxic if the organism is exposed to it twice as long. According to Suter (1993a), the model is equivalent to the hyperbolic function proposed by Green (1965), $LD_{50} = a + b(t^{-1})$, with the difference that there is no threshold, which makes it much more conservative.

Equation (1) has been used successfully by different authors, with the inevitable exceptions. For example, Anderson et al. (1981) showed that in prawn exposed to petroleum pollution, k is effectively a constant. They used the model in the form

$$LC_{50}\text{–d} = (LC_{50}\text{–96h}/4) \times d$$

where $LC_{50}\text{–d} = LC_{50}$ after d days of exposure.

There are more elaborate models, based on toxicokinetic constants that govern the flows of pollutants in organisms, such as the formula used by Kooijman (1981):

$$LC_{50} \infty = LC_{50t} [1 - \exp(-t/\tau)]$$

where τ is the speed of elimination constant, t is the time of exposure, $LC_{50\infty}$ is the LC_{50} for an infinite duration of exposure, and LC_{50t} is the LC_{50} for a duration of exposure equal to t.

The preceding models take into account a regular exposure, but the sources of pollution are often fluctuating. If these conditions cannot be reproduced in an assay, we must develop an appropriate model. One solution consists of measuring or modelling the source, graphically expressing the concentrations as a function of time, measuring the area under the curve, and deducing from it the mean concentration at each instant. In most risk evaluations of polluted sites, the exposure will be considered uniform during a long duration, except if cyclical or exceptional phenomena modify it (flooding of land, for example).

Linear combinations of plane models have been proposed to model, in three dimensions, the relations between dose, response, and effect. An example is the following function:

$$P = a + b \cdot Log(D) + c \cdot Log(t)$$

where P is the proportion of the population showing a response, D is the dose, and t is the time.

4.3.3. Acute and Chronic Toxicity

Acute toxicity assays are done in the short term, with high doses, around the LD(LC)$_{50}$, while, in most cases, ecological risk evaluation proposes to estimate concentrations without observed effect (NOEC) for prolonged exposures. It will therefore be advantageous to combine the extrapolation for the dose and the extrapolation for the duration of exposure in a single relation.

Different approaches have been used, notably regression and correlation models. Among the major contributions in this field, we must note the importance of the original data compiled by Slooff (1983), deZwart and Slooff (1983), and Slooff et al. (1983), because many methods of interspecific extrapolation, such as Kooijman's (1987), refer explicitly to these data. An equation enabling us to calculate the NOEC from the LC$_{50}$ (obtained from results of 164 assays on fish and daphnia) was developed by Slooff et al. (1986):

$$\log(NOEC) = 0.95 \, \log LC_{50} - 1.28 \ (r = 0.89)$$

The margin of uncertainty is 25.6 (the margin of uncertainty is defined by Slooff et al. (1986) as the value UF such that, the acute toxicity A of a product being known, the value of its chronic toxicity B is found within the values A/UF and A × UF with a probability of 95%). According to the authors, these results are close to those of Kenaga (1982), who estimated that 93% of the 135 ratios (log$_{acute\ toxicity}$/log$_{chronic\ toxicity}$) for 84 different products were equal to or less than 1.4.

The following formula, which enables us to calculate the MATC (for the life cycle of fish) from LC$_{50}$–96 h, was obtained from 31 combinations of species and chemical products (Andrews et al., 1977, in Suter, 1993a):

$$\log(MATC \cdot 10^5) = 4.78 + 0.675 \, \log(LC_{50} \cdot 10^2) \ (r = 0.679)$$

Other models were proposed by Suter et al. (1983, 1986) to calculate the MATCs from LC$_{50}$ for different species.

A method has recently been proposed to define three multiplicative factors of application enabling us to derive PNEC values from acute toxicity values, chronic toxicity values, or studies at the ecosystem level (ECETOC, 1993). Here we give an example of extrapolation from acute toxicity to chronic toxicity. To generate the corresponding factor of application, the geometric mean of different values of chronic toxicity (NOEC values) was calculated for each product, as well as the geometric mean of different values of acute toxicity (L(E)C$_{50}$ values), and their ratio (acute toxicity/chronic toxicity), based on published data for 360 substances tested on 120 aquatic species. The products were then classified in increasing order according to this ratio and the value of the ratio corresponding to the 90th percentile was determined. Different sets of data were tested, for example, all the species were considered, or just the freshwater species, or just fish, or just invertebrates, or both together (in each case, the number of products was variable, from 5 to 58; the metals were eliminated from the analysis). The results showed that in all cases, the value of the ratio is around 30, with one exception, where a value of 40 was found. In a conservative approach, it was decided that the value of the factor of application would be 40.

4.3.4. Nature of Product and Mix of Products

The quantitative structure–activity relationships (QSAR) serve to predict the toxic effects from the structure or physico-chemical properties of various congenerics (congenerics are defined as the set of molecules that have the same toxic effect but not necessarily the same mode of action, and for which it is possible to establish a continuous function linking the structure and the effect; Verhaar, 1994). The most famous example is the formula of Hansch et al. (1962):

$$\log(1/C) = a \cdot p + b \cdot \pi^2 + c \cdot \varepsilon + d \cdot S + e$$

where C is the dose corresponding to a determined effect, π is a term of hydrophobicity (for example $\log K_{ow}$), ε is a term depending on electronic characteristics of the molecule (for example the σ of Hammett), S is a term of steric congestion, p is product, d is dose, e is energy released, and a, b, and c are constants.

According to Verhaar (1994), one of the primary applications of modern QSARs in aquatic ecotoxicology was suggested by Konemann (1981), who showed that the LC_{50} of 50 industrial pollutants can be described fairly precisely by the following relation:

$$\log(1/LC_{50}) = 0.87 \cdot \log K_{ow} + 1.1$$

Most models used presently are simple mathematical relations, linking the LC_{50} (or the LD_{50}, NOEC, etc.) to the octanol–water partition coefficient[6] (K_{ow}). For example, the toxicity for freshwater fish (LC_{50}–96 h) of different substituted phenols can be estimated from K_{ow} with the following formula (Saarikowski and Viluksela, 1982):

---

[6]K_{ow} itself can be calculated from the chemical structure of the molecule and application programs are now available (for example, Devillers et al., 1995).

$$\log(1/LC_{50}) = 0.73 \cdot \log K_{ow} - 0.61 \text{ at pH } 6.0$$

or:

$$\log(1/LC_{50}) = 0.61 \cdot \log K_{ow} - 0.38 \text{ at pH } 7.0$$

These relations are extremely useful for predictive risk evaluation of new substances in the absence of toxicological data, but they are not useful for molecules belonging to homologous series that have a common mode of action. The study of QSAR is a rapidly developing field. Other types of relations have been proposed, but to examine them goes beyond the scope of this work (cf. Hermens, 1989; OCDE, 1992b, 1993a, b; MacCarty et al., 1993; Suter, 1993a; Donkin, 1994; Landis and Yu, 1995).

Mixtures of products are particularly difficult to analyse, but they occur in many polluted sites. There is the possibility of *synergy* (the effects of two toxins administered together are greater than the sum of the effects of two toxins administered separately; 1 + 1 = 5), *additivity* (the effects of two toxins administered together are equal to the sum of the effects of two toxins administered separately; 1 + 1 = 2), *potentiation* (the effects of one toxin are more intense in the presence of a non-toxic product (0 + 1 = 5), or *antagonism* (the effects of two toxins administered together are weaker than the sum of the effects of two toxins administered separately; 1 + 1 = 0.5), but these possibilities are poorly understood (Dybing, 1995; Seed et al., 1995; Landis and Yu, 1995).

In practice, the most frequent approach is either an additive approach or one taking into account the most toxic component of the mix. A variant of the additive approach is the use of TEQ (toxic equivalent), in which each compound is affected by a coefficient of balance (multiplicative) before the addition is made. This approach is conventionally used to evaluate the risk of families of products that have an identical mode of action, such as dioxins, PCB, and PAH. Extremely elaborate methods have been recently proposed to test the ecological effects of mixes and the associated uncertainties. These methods are based on the analysis of extrapolation error (cf. section 4.1; Logan and Wilson, 1995).

Bioassays are an excellent method of reducing the uncertainty associated with the 'monopollutant' character of the predictive approach.

4.3.5. Species: Generic and Specific Extrapolations

Toxicity assays are conducted with a limited number of species, which are not always those present in the ecosystem being evaluated. Taking these taxonomic differences into account corresponds to two different objectives of the evaluator, which caused Suter (1993a) to distinguish two types of interspecific extrapolation, the *generic taxonomic extrapolations* and the *specific taxonomic extrapolations*. This distinction does not correspond to the difference of hierarchical level between the genus and the species, but to two different objectives of the process of extrapolation.

Specific extrapolations aim to estimate the toxic effects on a species from assays done on another species, for example, between rat and human, trout and perch or carp, Japanese quail and partridge or kestrel, or even, considering larger taxonomic differences, between invertebrates and vertebrates. We can sometimes give qualitative, intuitive responses to this question: for example, insects will be inherently more sensitive to insecticides than vertebrates. Specific extrapolation is not always judged to be necessary (ECETOC, 1993).

Generic extrapolations use toxicity data on one or several species to estimate the distribution of toxicities in a community (e.g., the values of different LD_{50} for different species). We will return later (section 4.4) to generic extrapolation in reference to its possible use in regulatory strategies.

The simplest method for estimating the toxicity of a product for a species other than that of the assay is to use a factor of correction, which will be the ratio of toxicities of other products for these two species. For example, if species A is twice as sensitive as species B to product X, we estimate that it has strong chances of being twice as sensitive to product Y.

$$LC_{50}A,Y = LC_{50}A,X \cdot (LC_{50}B,Y/ LC_{50}B,X)$$

where $LC_{50}A,Y$ is the LC_{50} of product A for species Y, $LC_{50}A,X$ is the LC_{50} of product A for species X, $LC_{50}B,Y$ is the LC_{50} of product B for species Y, and $LC_{50}B,X$ is the LC_{50} of product B for species X.

The value of this type of equation admittedly depends on chemical structures and the taxonomic proximity of species. The closer the species, and the closer the chemical structures of the pollutants, the better the estimation will be.

Another technique of interspecific extrapolation is the dose-scaling widely used in health risk evaluation but less so in ecological risk evaluation. Dose-scaling consists of transforming the toxic doses by a factor of correction, which then enables us to directly compare the toxicities, in other words to express the results on the basis of a common unit. The expression of results in mg/kg live weight is a classic application of this technique, but other modes of expressing toxicity can be more effective in 'removing' interspecific differences. For example:
- the dose per unit of surface area;
- allometric regression;
- lipid content.

For example, Patin (1982) finds a strong correlation between the logarithm of body length of various aquatic species and the toxicity (LC_{50}) of several organic and inorganic pollutants. The correlation coefficients for each pollutant are relatively high (0.73–0.91), although the assays were done with different types of organisms (from microorganisms to fish), the test conditions were different, and the sizes were sometimes estimated and not measured. The lipid contents could standardize the toxicities of lipophilic pollutants (Geyer et al., 1993). Figure 3.1 shows the relation between body weight and toxicity of methiocarb (carbamate insecticide) for birds (Mineau et al., 1996).

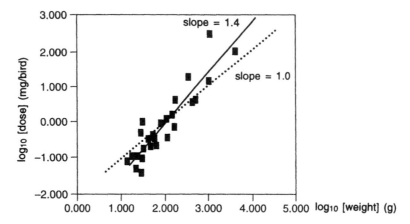

Fig. 3.1. Example of a statistical regression model to estimate the toxicity of a product for two different species (according to Mineau et al., 1996; with the permission of Academic Press, 1997). The squares represent the LD_{50} values of methiocarb (carbamate insecticide) for different bird species. The solid line represents the line of regression based on a relation of the type $\log(LD_{50}) = \log(a) + b \cdot \log(\text{body weight})$. The broken line represents the relation $\log(LD_{50}) = \log(a') + \log(\text{body weight})$.

Some biological factors can take into account taxonomic differences such as the feed regime. Polyphagous organisms will be less sensitive to toxins than oligophagous ones, which in turn are less sensitive than monophagous ones (Croft and Morse, 1979). These results indicate only the trends and, therefore, are not easy to use in risk evaluation.

General methods can be adapted to particular cases. For example, the approach used in the Netherlands to define the maximum acceptable concentration of a substance in the water or soil (MAR,[7] maximum acceptable risk level) is based on NOEC values for aquatic species. It is not adapted to risk of biomagnification in certain species that are particularly at risk,

---

[7]This value corresponds to HC5, that is, the concentration without effect, calculated by extrapolation and supposed to protect 95% of the species in an ecosystem (cf. section 4.4.2).

such as fish-eating birds. To take this risk into account, Romijn et al. (1993) proposed the following algorithm (cf. chapter 2, section 4.3.3):

$$MAR_{water} = NOEC_{feed} \times BCF^{-1}$$

where $NOEC_{feed}$ is the NOEC value determined by administration of a product in the feed and BCF is the bioconcentration factor.

However, the NOEC values used in this formula are based on concentrations of a substance in the feed (commercial) given to domestic animals used for ecotoxicity assays and do not take into account certain characteristics of natural exposure of marine birds:

- Composition of feed: the caloric value (kJ/kg) is not the same for feed granules, invertebrates, and fish.
- The energy needs of the species concerned are not identical: the level corresponding to caged birds (existence metabolic rate, EMR) is different from that of birds in natural conditions (field metabolic rate, FMR). The latter can be divided into two components, according to whether conditions are normal or extreme (moulting, reproduction, migration, or very unfavourable climatic conditions).

A factor of correction was recently proposed to take these differences into account (Everts et al., 1993). The $NOEC_{natural feed}$ (dose without effect in the natural feed) is derived from the $NOEC_{test feed}$ (dose without effect in the feed used in the test) by the following formula:

$$NOEC_{natural feed} = NOEC_{test feed} \cdot (E_{prey}/E_{test feed}) \cdot$$

$$(EMR_{test-bird}/FMR_{natural bird}) \cdot (FMR_{normal}/FMR_{maximum})$$

where $E_{prey}/E_{test feed}$ is the ratio of the caloric value of the 'natural feed' (for the 'natural bird', that is, the species at risk) to the 'test feed'. The values of this ratio for different types of feed (insects, molluscs, several species of fish) are found in the reference indicated. The ratio $(EMR_{test-bird}/FMR_{natural bird}) \cdot (FMR_{normal}/FMR_{maximum}$ will be 2.75 if the tests are done on Galliform adults and if the species at risk belong to the family Charadriidae (plovers). The energy needs can be adapted to the weight of the animal by a formula of the following type:

$$logEMR = a + log(body weight)$$

The MAR_{water} is calculated finally by the formula:

$$MAR_{water} = NOEC_{natural feed} \times BCF^{-1}$$

Most of the taxonomic extrapolations were developed for aquatic animals and, in some cases, for vascular plants (Fletcher et al., 1990).

4.3.6. Development Stage, Size and Age

For any species, the size and development stage of animals influence toxicity. The juvenile stages are generally more sensitive than the adult for several reasons:
- Greater exposure: juveniles have greater body surface (exposure by cutaneous route), oxygen needs (exposure by pulmonary route), and nutrition needs (exposure by intestinal route) than adults.
- Less active metabolism systems than adults.
- More sensitive targets than in adults (growing organs).

The contribution of these different factors to toxicity is highly variable. With regard to acute toxicity, fish eggs are often considered the least sensitive stage and alevins the most sensitive, the sensitivity of adults being intermediate. In aquatic invertebrates, sensitivity to toxins diminishes with age. The published data show that sensitivity of adult birds to pesticides varies very little with age.

Extrapolation based on size has already been seen with dose-scaling, but within the area of interspecific extrapolation. The techniques used are effective for intraspecific extrapolation as well, obviously with a smaller margin of uncertainty.

4.3.7. Tolerance, Adaptation and Resistance

Adaptation is reduction in sensitivity to stress in an animal or plant species, resulting from biochemical or physiological compensation of the organism in the face of environmental conditions. *Resistance* is reduced sensitivity of certain populations, depending on hereditary factors that are genetically transferable. The most remarkable example is the resistance of certain populations of insects to insecticides. *Tolerance* is the lack of innate sensitivity of a species in comparison to another. These expressions are very often used interchangeably. Resistance and adaptation are rarely taken into account in most ecological risk evaluations.

4.3.8. Mode of Exposure

The avenues of exposure used in toxicological assays are not necessarily those that exist in nature. For example, a test protocol may be designed to determine the toxicity of an insecticide to a bird after intravenous injection, which is obviously very different from natural conditions, in which the alimentary route is one of the chief avenues of exposure. In fact, the protocols attempt to approach natural conditions as closely as possible and, in practice, a toxic product is never tested on birds by intravenous injection, but is administered orally in a capsule. However, the quantities administered to the animal in the test (mg of toxin) must be converted into quantities consumed in natural conditions. This conversion is easily done by multiplying the concentration in the feed by the (estimated) quantity consumed and adding the alimentary loads, but it brings into play a series of hypotheses that will increase the uncertainty of the result.

- the bioavailability of the product is the same under all circumstances;
- the product does not have a repellent effect;
- the product is stable in the feed;
- the durations of administration of the product must not be too different, otherwise a short-term test has to be extrapolated to a long-term test.

A classic example of differences in modes of exposure is that which exists between the static and dynamic assays for aquatic fauna. Mayer et al. (1986) analysed the results of 123 pairs of static and dynamic tests for 41 chemical products and showed that this ratio varies from 0.12 to 8.33 with a mean of 0.51. The value of 0.51 does not seem advisable to extrapolate from one type of test to another, but a factor 10 (conservative value) can be proposed to take into account differences between the two types of test.

4.3.9. Nature of the Environment

Extrapolation of the nature of the environment is designed to take into account differences between environmental conditions and test conditions in, for example, soil composition or water. Temperature is a significant cause of variation in toxicity for poikilothermic species. Most products are more toxic when the temperature rises, but there are important exceptions; for example, the toxicity of DDT for invertebrates increases when the temperature falls. Variations in sensitivity of individuals in relation to cyclical physiological variations (annual cycle) are another cause of variation in toxicity. Correction factors can be applied to take these phenomena into account, factors that can be calculated from parametric tests (cf. chapter 1, section 2). Let us suppose, to take again the example of DDT, that we wish to calculate the risk for invertebrate soil fauna at 2°C, with results of ecotoxicity assays done at 15°C. We search the literature to see whether toxicity values at different temperatures are available for different invertebrate soil species, and we then calculate by regression the rate of reduction in toxicity by degree and deduce an empirical correction factor to evaluate risk at a given temperature.

4.4. Extrapolation to Populations and Communities from Monospecific Assays

It is difficult to establish standards for the quality of mediums from results of an ecotoxicity assay using a single species. Intuitively, it seems more realistic to base it on a greater number of assays incorporating species from different taxonomic levels (a battery of tests). The weakest points of the system (the most sensitive species) are detected more easily and the standards are thus ecologically pertinent. Sophisticated methods have been elaborated to analyse the data collected on several populations and attempt to simulate the reality of ecosystems, such as the model proposed by van Straalen and Denneman (1989). However, the juxtaposition of studies on different populations does not seem sufficient to ensure ecological relevance, and other approaches have been proposed.

4.4.1. Criteria for Selection of Tests and Batteries of Tests

The characteristics required for a 'good' assay species are numerous and often contradictory. According to the terms of an exhaustive analysis, Leon and van Gestel (1994) proposed a series of criteria that can be used for the selection of ecotoxicity assays. These criteria are divided into two groups: criteria for selection of monospecific assays and criteria for selection of a battery of ecotoxicity tests (Table 3.2).

Criteria for selection of monospecific assays

The *practicality characteristics* are feasibility, cost-quality ratio, and rapidity. According to the authors, *feasibility* corresponds to the ease with which

Table 3.2. Criteria used by Leon and van Gestel (1994) to select tests and batteries of tests of terrestrial ecotoxicity

---

Criteria for selection of monospecific tests

1. Practicality
   a. feasibility
   b. cost-quality ratio
   c. rapidity
2. Acceptability
   a. standardization
   b. reproducibility
   c. statistical validity
   d. good laboratory practices
   e. extensive response to pollutants
3. Ecological relevance
   a. ecological realism
   b. biological validity

Criteria for selection of batteries of tests

4. Representativity of terrestrial ecosystem
   a. representativity of adaptive strategies
   b. representativity of functional groups
   c. representativity of taxonomic groups
   d. representativity of avenues of exposure
5. Representativity of responses for the terrestrial ecosystem
6. Uniformity

---

the assay organisms are obtained, maintained, cultivated, and handled. To that can be added the possibility of completing the assay with relatively unsophisticated means. The *cost-quality ratio* covers three elements: the cost of installations, the cost of the test itself, and the cost of personnel. A compromise must be made between the cost and precision of the assay. If the number of individuals is increased, the precision increases, along with the cost. The *rapidity* of the assay depends on duration of exposure, duration of the post-exposure period (which must be sufficient to allow the toxic effect to manifest), and duration of the observation period. For example, it would be difficult to test toxic effects over a biological cycle if that cycle were long.

Acceptability refers to standardization, statistical validity, good laboratory practices, and the capacity to respond to several pollutants.

Standardization aims to:
- facilitate the completion of the assay with the help of an experimental protocol;
- facilitate the planning of the assay;
- minimize the probability that the assay protocol will be misinterpreted;
- increase the compatibility of the results;
- enable a classification of products according to the danger they present.

This standardization is done at various levels:
- methods of raising organisms;
- analytical methods;
- statistical methods;
- environmental conditions (temperature, atmospheric humidity, availability of food resources);
- characteristics of the medium (for soil, these include pH, organic matter content, structure, and relative humidity);
- characteristics of individuals (age, size, developmental stage, origin);
- characteristics of exposure (means, vehicle, frequency, duration, concentration).

Reproducibility represents the possibility of replicating an assay. Even with very detailed assay protocols and rigorously sorted organisms, there remains some variability in the results. That variability can be tested by inter-laboratory calibrations. This is an important stage in the risk evaluation procedures to ensure the quality of data. *Statistical validity* supposes a random distribution of individuals in the different lots treated and a correct treatment of data by the appropriate statistical methods (Gelber et al., 1985; Hoekstra, 1991). Good laboratory practices are defined as a scientific and management concept covering the organization and conditions in which the laboratory studies are conceived, completed, followed up, recorded, and reported. The *capacity to respond to numerous pollutants* signifies that the test will be sensitive to a range of products with similar characteristics, for example, all insecticides.

Ecological relevance of a test is one of the points that receive the most discussion. *Ecological realism* covers several aspects[8]:

- The test species, at the development stage at which it is tested, must play a functional role in the community.
- The test species must be a good representative of the taxonomic group to which it belongs, that is, it must not be too different from other species of the same group in its morphological, physiological, and ecological characteristics. From this point of view, cosmopolitan species are more useful than endemic species.
- The final points must be ecologically relevant, for example, survival rate, growth, mobility, nutrition, and reproduction.
- If reproduction is measured as a final point, the duration of exposure must cover the major part of the period of sexual activity of the species. If growth is measured, the exposure must cover the entire period of maturation (for species of finite growth) or most of it (for species of continuous growth).
- The biotic and abiotic factors must simulate the natural habitats as well as possible. This is particularly difficult to do for soils, taking into account variations in structure, texture, and environmental conditions (humidity, temperature).
- The avenues of exposure and bioavailability of the chemical product must be identical to those that occur in natural conditions.
- The concentrations of a product and the duration and frequency of the exposure must also be similar to natural situations. Unlike in aquatic ecotoxicology tests, it is difficult to change the medium in assays on soil organisms. The concentrations indicated will be the nominal concentrations.

Biological validity signifies that the test results must be attributed solely to the influence of the toxin and not to other factors. For this purpose, the test must meet two conditions:

- it must begin with healthy animals;
- the survival and reproduction of the control group must meet pre-established conditions.

DeGraeve et al. (1992) defined three criteria to ensure that the assay organisms are in good health at the beginning of the assay: quality of survival of parental organisms, quality of their reproduction, and defined

---

[8]Under the phrase 'for reasons that are not evident', Leon and van Gestel group several characteristics of a test: *sensitivity*, defined as the capacity to respond to low doses (for the same living element, the sensitivity of a test depends on the duration of exposure and the dose, and it can be evaluated by the value of the final point), and the capacity to provide *exact results*, without false positives or negatives (this point corresponds to *biological validity*).

age (or development stage). Measuring the response to a reference toxin is an excellent practice to characterize the sensitivity of the stock population and the response can be considered a criterion of its good health.

Criteria for selection of batteries of tests

Criteria were established for selection of batteries of tests because of the need for a balanced group of representative species, not only of different taxonomic groups, but also of various trophic levels, adaptive strategies, habitats, etc.

The battery must include *species representing different adaptive strategies,* r strategies and K strategies, for example.[9] It can be assumed that populations with K strategies will have greater difficulties in repopulating contaminated mediums than populations with r strategies. The use of rare species that are in danger in the ecosystem at risk would be interesting because it would be relevant to the problem to be resolved, but such species are often protected by national and international laws, justifiably, to prevent significant extractions that may decrease their numbers.

The *principal functional groups* of ecosystems must be represented, because the evaluation of effects must protect the entire ecosystem. This parameter is optimized by taking species belonging to different trophic levels. However, an organism's position on one particular trophic level (herbivore, decomposer, etc.) is not sufficient to characterize its membership in a functional group because the resources exploited may be different (all herbivores do not consume the same plants). Leon and van Gestel did not directly refer to the possibility of bioaccumulation in the food chain, but we must allow for it.

The representation of *different taxonomic groups* is necessary, in the initial hypothesis of protecting the entire ecosystem (Table 3.3). Leon and van Gestel mention also under this heading the need to take into account the

---

[9]The following points are taken from Barbault (1992, 1993), in which the interested reader can find detailed explanations. The definitions of r and K strategies correspond to two different types of biodemographic strategies. These strategies are defined on the basis of the logistic equation of Verhulst, which takes into account the numerical growth of natural populations:

$$N_t = N_{t-1} + r_m(1 - N_{t-1}/K)N_{t-1}$$

This expression signifies that the population count at the end of the interval considered (N_t) depends on the initial number (N_{t-1}), on r_m, the maximum value of r (rate of growth per individual: $r = (N_t - N_{t-1})N_{t-1}$), and on K, the equilibrium density. From this model, MacArthur distinguished two types of selection:
• Type r, which occurs in populations of low density in the process of expanding, tends to promote the highest possible rate of reproduction (maximization of r) and favours r genotypes, that is, species with a short cycle of generation and high rate of reproduction.
• Type K, which occurs in dense, stationary populations, tends to promote a better conversion of trophic resources to descendants (maximization of K) and favours K genotypes, that is, species with long life-span and low rate of reproduction.

Table 3.3. Taxonomic groups representative of terrestrial ecosystems (according to Leon and van Gestel, 1994)

bacteria	molluscs
actinomycetes	arachnids
fungi	isopods
algae	myriapods
vascular plants	insects
protozoa	poikilothermic vertebrates
nematodes	homeothermic vertebrates
annelids	

diversity of habitats, especially to incorporate epigaeal and endogaeal species in the battery of tests (but that may constitute a separate heading). Another reason to incorporate different species is the hypothesis (which remains to be proved) that species of the same taxonomic group have sensitivities to toxins that are overall more similar than those of species belonging to very different taxonomic groups.

The last point concerns the selection of organisms *exposed by different routes* in contaminated mediums: air, soil solution, feed, solid soil particles.

Leon and van Gestel do not provide precise criteria to determine point 5 (cf. Table 3.4), the *representativity* of biological responses for the ecosystem, and rely only on two hypotheses:

- If different species of the terrestrial ecosystem are protected, their interactions will be preserved and, consequently, the functioning of the ecosystem will also be preserved.
- If representative test species are selected, the statistical methods of extrapolation will serve to forecast the responses of the ecosystem.

These hypotheses, it must be noted, have not been verified and have been criticized by other ecotoxicologists (see the following section).

Table 3.4. Functional groups representative of terrestrial ecosystems (according to Leon and van Gestel, 1994)

primary producers
decomposer micro-organisms
detritivorous invertebrates
other invertebrates
pollinators
parasites and parasitoids
herbivorous invertebrates
herbivorous vertebrates
predatory invertebrates
predatory birds and mammals

The final criterion that a battery of tests must satisfy is *uniformity*, which consists, in the logic of standardized tests, of having the least variation of factors possible. For example, the same type of soil and plant is used in each test.

4.4.2. Methods of Generic Extrapolation: Distributions of Sensitivity

No matter how much care is taken in the selection of criteria, it remains true that the number of species tested is very small in comparison to the millions of species existing. One of the principal problems of ecological risk evaluation is the transposition of information obtained from standardized ecotoxicity assays done on some species to natural populations and communities.

A primary solution consists in retaining the lowest of different known values, for example, if one knows the NOEC of cadmium for several species of soil invertebrates, one holds the lowest (possibly after application of a factor of security) as a criterion of absence of danger of cadmium-contaminated soil; that is, one retains the *most sensitive species.*[10]

Regression models, enabling the extrapolation of data from monospecific ecotoxicity assays to 'ecosystems', were proposed by Slooff et al. (1986). These authors used different values of toxicity of 34 chemical products for aquatic fauna (values reported in the literature): acute EC_{50}, chronic NOEC ($NOEC_{monospecific}$), and toxicity in conditions representative of an ecosystem ($NOEC_{'ecosystem'}$, in the terminology used by the authors, corresponds to a multispecific system, a microcosm, or field assays). They deduced from it an equation enabling the calculation of $NOEC_{ecosystem}$ from $LC(E)_{50}$:

(1) $\log NOEC_{ecosystem} = -0.55 + 0.81 \cdot \log LC(E)_{50}$ $(r = 0.77)$

with a margin of uncertainty (cf. section 4.3.3) of 85.7. Another equation links $NOEC_{ecosystem}$ to $NOEC_{monospecific}$:

(2) $\log NOEC_{ecosystem} = 0.63 + 0.85 \cdot \log NOEC_{monospecific}$ $(r = 0.85)$

with a margin of uncertainty of 33.5. However, taking into account the relative imprecision of data on ecosystems, the authors modified these data by using factors of application of 0.1 and 10, respectively, for the lowest and highest values, and then deduced two other equations:

(1a) $\log NOEC_{ecosystem} = -0.84 + 0.73 \cdot \log LC(E)_{50}$ $(r = 0.72)$

with a margin of uncertainty of 106.6 and:

---

[10]That is, the most sensitive species in the assays taken into account for the evaluation, but that does not mean that this species is the most sensitive of all existing species.

(2a) $logNOEC_{ecosystem} = 0.30 + 0.80 \cdot logNOEC_{monospecific}$ $(r = 0.80)$

with a margin of uncertainty of 47.0. In the opinion of the authors, these results indicate that it is not impossible to estimate the long-term effects from monospecific ecotoxicity assays.

Other methods have been developed to estimate a threshold of absence of toxicity other than by the choice of the lowest value. Most are based on the concept of distribution of sensitivities developed essentially by Kooijman (1987). The principle of it is simple: the set of LC_{50} values (or other final points, such as NOEC) of a number n of species for a given substance is described by a histogram and then a statistical distribution. This distribution can be considered the distribution of sensitivities to the toxin for all the populations. From these principles, Kooijman (1987) proposed a model based on a logistic distribution of log of LC_{50} values, designed to protect the *most sensitive* species in the community. Kooijman's model was criticized as being too conservative and another model developed on the same principles was proposed by Van Straalen and Denneman (1989). The value calculated after application of this model (HC5) is supposed to protect 95% of the species of the community. Other analogous methods were proposed (Wagner and Lokke, 1991; Aldenberg and Slob, 1993; Emans et al., 1993). These methods are very important because they can serve as the basis for regulation of the admissible concentrations in the mediums (van Leeuwen, 1990; DGM, 1990). Descriptions with further details can be found in Annexure 3.

4.4.3. *Criticism of Methods of Generic Extrapolation*

Extrapolation models have been very much criticized, notably by the team of Cairns (Smith and Cairns, 1993). According to these authors, the general principle of these methods is the following:

- To choose a final point: usually, the LC_{50} or NOEC, for which a large number of data are available, but also other values, such as LC_{10} or data of another kind.
- To postulate a distribution for the final point or a transformed value: for example, a logarithmic transformation of data to take into account bias in the distribution. The log(final point) is supposed to follow a normal distribution (Wagner and Lokke, 1991), a logistic one (Aldenberg and Slob, 1993), or a triangular one (Stephan et al., 1985).
- To collect data for a number n of species and estimate the dose without effect from the formula $log(HC\alpha, \gamma) = X - K\alpha, \gamma \cdot s$, where $HC\alpha, \gamma$ is the concentration that protects $100(1 - \alpha)\%$ of the species with an interval of confidence of $100 \gamma \%$, X is the mean of the data, s is the standard difference, and $K\alpha, \gamma$ is a factor of adjustment based on α, γ and n.

For the extrapolation model to function correctly, some statistical and biological hypotheses are needed.

First hypothesis: the final point is a sample in the distribution

The empirical data compiled for a large number of species give a symmetrical histogram of values of log(final point). For Smith and Cairns, the different types of distribution used have not been truly verified with small samples, but it is hardly probable that the type of distribution used (normal, logistic, or other) will be very important in relation to other hypotheses formulated. In practice, the three distributions mentioned give very similar results. The Kooijman model and subsequent ones imply necessarily that the populations of each species follow the same distribution and that each specific final point must be considered as an element of this distribution, which supposes for each species the same type of distribution and the same variance. In fact, the final points are estimated by different studies the results of which are affected by the size of the sample and the test protocol. The variation of values of final points can be different for the same theoretic value. A theoretical and practical problem therefore arises: how can we model the data with different variations?

Second hypothesis: extrapolation works better if the distribution of sensitivity in populations of test species is the same as those of natural populations

This is the hypothesis of 'representativity'. There is little chance of verifying this hypothesis precisely because test populations of species are selected to reduce the heterogeneity of individual variations.

Third hypothesis: test conditions in the laboratory do not significantly influence the sensitivity of species to the toxin

The standardized protocols are useful to ensure uniformity and maintain reproducibility of results of assays over time and between laboratories. However, there is variability in results obtained with the protocols, even the best standardized, such as daphnia tests. Two approaches are possible to take factors of confusion into account either to develop more complex models of sensitivity or to use regression models to adjust the measurement of final point at the factor of confusion. In fact, most of the parameters responsible for much individual variability are already identified and incorporated in the assay protocols (for example, for an aquatic ecotoxicity assay, water temperature will be fixed) and the differences that persist in the inter-laboratory calibrations pose a problem, justifiably, because one does not know whether they originate in the animal or the laboratory.

Fourth hypothesis: the sample is a random sample

This hypothesis is necessary to estimate the mean and variance of the distribution and deduce the percentile (Hi_p). The laboratory species are not chosen randomly, but for reasons of convenience, cost, or economic importance. Most works concern species said to be 'sensitive', but without proof of particular sensitivity in comparison to other species. For example, of the 769 North American freshwater fish species, the toxicity of only a single pollutant has been tested on just 6%, and the toxicity of a very large number of pollutants has been tested only on a few of them, mostly salmonids (Suter, 1993a).

Fifth hypothesis: the final point measured is relevant

Certain final points pose statistical and biological problems. The value of NOEC, for example, depends on the size of the sample. Small samples tend to give higher NOEC values with different variances. Smith and Cairns remark that if one uses LC_{50} rather than other LC levels, the values obtained are affected with less uncertainty because the variance is generally smaller around the log-dose integer curve, that is, around the LC_{50}. Moreover, to protect 95% of the species from values that define the mortality of 50% of animals can be judged unacceptable. The use of a factor of adjustment may partly solve this problem.

Sixth hypothesis: the concentration limit based on a certain final point protects higher levels of organization

The use of batteries of tests and methods of generic extrapolation presupposes that the populations and communities are independent elements. The criticism leveled by Smith and Cairns concerns chiefly the prediction of effects that are properly speaking ecological, that is, relations between species, such as prey–predator relations, or competition between species for food or habitat, which are not obviously predictable from monospecific ecotoxicity assays. Another major problem is that effects other than mortality can disturb the equilibrium of populations, such as diminished reproduction, which are difficult to predict from LD_{50} values. Finally, taking again the data of Slooff et al. (1986), they emphasize the distinction to be made between a good statistical coefficient of correlation and a good predictive value.

Seventh hypothesis: the other approaches are costlier and do not provide supplementary information

Cairns and Smith's stand brings to light an important element in risk evaluation that is particularly difficult to assess: its cost. It is part of the evaluation itself (which is the best approach for a defined cost) and depends also on risk management (knowledge has a price: the greater the means, the better the results).

Eighth hypothesis: the method provides values that are too conservative

The demonstration of Smith and Cairns, based on the analysis of data of Slooff et al. (1986), shows that this approach provides effectively conservative values, even for a small number of data, but only if the other hypotheses are verified.

Real value of methods of generic extrapolation

For Smith and Cairns, the information given by ecotoxicity assays enables us to predict risks for the laboratory population of the species. However, despite the criticisms of standardized assays, they admit that evaluations have long been based on this approach. The case of polluted soils and sites is peculiar. If one wishes to develop standards of toxicity for *polluted soils*, it is probable that danger evaluations and fixation of limits will be the best approach, whereas risk evaluations of *polluted sites* benefit from information provided by *in situ* indicators and bioassays.

4.4.4. *Other Approaches*

Extrapolation is not the only means of modelling effects at the level of populations and communities, possibly ecosystems, from monospecific assays. The development of *ecological models* has considerable theoretical advantage, even if the practical applications are still very limited. This approach consists of integrating the toxicity data in ecological models describing the structure and functioning of populations (Barnthouse, 1992; DeAngelis et al., 1990; DeAngelis and Gross, 1992).

The final points traditionally used in ecology, such as abundance of populations, biomass, and size and age distributions, can be used for this purpose. Ecologists have long elaborated dynamic models of populations as a function of variables such as processes of recruitment (birth rate and immigration) and processes of disappearance (mortality and emigration).

Some models are used by resource managers, for example pisciculturists, to evaluate the impact of fish captures on the survival of populations. Barnthouse (1992) remarks that they can be applied to acute toxic effects of chemical treatments if mortality is considered a particular form of yield. Other models hold great interest, for example, those that estimate the probability of extinction of a species over a given period, as a function of population size or necessary habitat (Barnthouse, 1992).

Predictive models at the level of the ecosystem, essentially aquatic, have been proposed. These models integrate abiotic variables (temperature, illumination, feed) and biotic variables (e.g., toxicity data, population parameters at different trophic levels). Suter (1993a) summarized the characteristics of existing models and models that can be used in ecological risk, considering minor modifications. He classifies these models under three major headings:

- *Energy models*, based on the analysis of energy flows in the ecosystems (productivity). These are the most numerous.
- *Models of biogeochemical cycles* (e.g., carbon cycles, nitrogen cycles, and sulphur cycles in a prairie: SAGE model, Heasley et al., 1981).
- *'Naturalistic' models*, that is, according to Suter, models of structure of the trophic chain (e.g., number of trophic relations, species wealth, proportion of primary consumers).

Suter remarks that very few models of this type are operational in ecological risk evaluation. One of the chief obstacles to development of this approach lies in the absence of clearly defined final points adopted by the scientific community and regulatory authorities.

The most recent models of population dynamics have been used in environmental toxicology. The Verhulst equation can lead to different types of evolution of populations according to the value of r (cf. section 4.4.1): equilibrium, stable fluctuations, or chaotic dynamics. The use of non-linear dynamics in environmental toxicology is presented in the work of Landis and Yu (1995). A very interesting introduction to the application of multivariate analysis to biological systems disturbed by pollutants (microcosms, multispecific systems) can be found in the same work.

Most models can be used interchangeably in retrospective and predictive evaluation. In a retrospective evaluation (polluted sites), a value of exposure is derived that corresponds to the threshold of acceptable effect from a model of effects. This exposure value then serves to calculate the levels of environmental concentrations without danger, for example, for the rehabilitation of a site (Fordham and Reagan, 1991).

5. INTEGRATED ASSAYS

There is a gap between laboratory assays and natural situations, in terms of dimensions and ecological realism. It is therefore tempting to use more complex physical models, with the idea of approaching the conditions of the natural medium. The disadvantages of these systems have often been mentioned, especially their low reproducibility and their very high cost in comparison to monospecific assays. Their use in risk evaluations remains very limited, but is increasing. The chief advantages of these systems are the following:

- study of biological phenomena on the larger spatial (possibly temporal) scale, in environmental conditions closer to those of the natural medium than with ecotoxicity tests;
- study of phenomena at higher levels of biological organization; and

- validity of simulation models.

In order of increasing size and complexity, the following can be distinguished:
- multispecific laboratory assays;
- experimental trophic chains;
- microcosms; and
- mesocosms.

Multispecific assays on communities of aquatic micro-organisms have been particularly studied by Cairns et al. (1992). Their development for regulatory purposes of evaluation has been debated by Cairns (1988) and LaPoint and Perry (1989).

Manipulated ecosystems, enclosures, and *impact studies* will be considered not here, but in the section on *in situ* indicators. In fact, there are no distinct divisions or oppositions between these categories, simply an evolution towards increasingly complex models. Even natural situations can be considered 'life-size' experiments, just as experimental assays are 'microcosms' in the etymological sense of the term (cf. chapter 1, section 2.3).

SETAC defined *microcosms* as systems of volume less than 15 m^3 or less than 15 m long, which distinguishes them from *mesocosms,* but the principles that regulate these two categories are fundamentally the same.

Microcosms were highlighted and studied in the 1970s, particularly by the team of Metcalf (1971), under the name 'model ecosystems', to examine the occurrence and effects of fertilizers and pesticides. These microcosms, mostly aquatic systems, use solid mediums (sediments), liquid mediums, and living elements in contact with each other in closed or open systems (which could take into account exchanges with the atmosphere).

Terrestrial microcosms were introduced much later. According to Morgan and Knacker (1994), a terrestrial microcosm is a controlled, reproducible system that attempts to simulate the functioning of a part of the terrestrial ecosystem. In this system, environmental factors such as light, relative humidity, and precipitation are controlled. The microcosm must encompass more than one trophic level. Here we distinguish between the study of a plant in a pot growing on polluted soil (bioassay) and the study of a medium made up of this plant and soil micro-organisms, the complete system making up a true microcosm.

The classification proposed by Morgan and Knacker (1994) divides terrestrial microcosms according to two properties:
- the integrity of the terrestrial medium (carrots of soil extracted and used as such or homogenized soil);
- the degree of openness of the aerial medium (open or closed system).

Other details on terrestrial ecosystems can be found in the article by Verhoef and van Gestel (1995).

The advantage of developing microcosm and mesocosm approaches has been discussed by Gillett (1989) and Cairns et al. (1993a). Some of these systems have been standardized by the EPA. The Soil Core Microcosm (SCM) was developed by the ASTM (1987) to test xenobiotics in agro-ecosystems and details on aquatic systems can be found in other publications (SETAC, 1991a, b). The microcosmic approach can probably play a greater role in evaluation of polluted sites. Microcosms can be constructed from polluted mediums (water, sediments, and soil) in a bioassay type of approach. The mesocosmic approach has inherently less advantage.

6. *IN SITU* INDICATORS OF EFFECTS

The response of indicators of effects can be measured *in situ* in *enclosures, natural ecosystems*, or *manipulated ecosystems*. The enclosures are isolated parts of ecosystems. Manipulated ecosystems are natural ecosystems in which one or several characteristics have been artificially modified. One well-known example is Weber lake (Wisconsin, USA), enriched with a mix of organic and inorganic fertilizers (Juday et al., 1938). An ecosystem can be manipulated in different ways:
- introduction of pollutant (impact studies);
- modification of habitat (artificial substrates);
- introduction of living organisms (systems of enclosure).

A review of manipulated ecosystems can be found in the article by Evans and Dillon (1995).

6.1. General Principles

Risk evaluation from ecotoxicity assays is a predictive method that is hard to pursue beyond the level of the individual. It can be generally admitted that extrapolations or the use of integrated systems ensures a better estimate of effects on populations and ecosystems, but the former involve a high degree of uncertainty and the latter is expensive. Whatever the method used, these estimations can be modified, in one way or another, by *in situ* studies. The case of polluted sites is interesting, because they are 'life-size experiments' that can provide a large amount of information on exposure and toxic effects on local populations, but the medium (soil) is the most difficult to study and exposure diagrams are very complex because of the frequent presence of many pollutants.

Field data provide numerous advantages (Warren-Hicks et al., 1989):
- Autochthonous species are continuous indicators of environmental pollution, integrating short-term fluctuations.
- The effects are directly measured on individuals belonging to natural populations, so extrapolations are not needed.
- Results can be directly interpreted, since they concern the populations at risk.

- Data on autochthonous populations are much better understood and accepted by local managers and the public than more theoretical data on species known only to a few specialists.
- Field data allow researchers subsequently to ensure the efficacy of site rehabilitation measures.
- In the diagram of Warren-Hicks et al. (1989), field data constitute an indispensable stage to complement chemical and toxicological data and obtain an integrated risk evaluation of polluted sites.

The *in situ* approach, however, has some disadvantages:

- All the populations cannot be studied. It is difficult to study the impact of a polluted site with small surface area on populations of large mammals or predators that come to the site occasionally. Generally, populations and communities of terrestrial or aquatic bacteria or invertebrates are studied.
- All the methods are not standardized (even though the methodologies are recognized by the scientific community).

The terminology used in this field is very fluid. It has frequently been defined by ecologists for the purposes of bio-evaluation ('procedures with biological basis used in establishing ecological diagnostics'; Blandin, 1986) or biosurveillance (follow-up).

A *biological indicator* (or bioindicator) is defined by Blandin (1986) as 'an organism or set of organisms that, with reference to biochemical, cytological, physiological, ethological, or ecological variables, practically and with some certainty enables the characterization of the state of an ecosystem or an ecocomplex and an indication as early as possible of their natural or provoked modifications.' For Ramade (1993), 'biological indicators of pollution are species particularly sensitive to a contaminant that serve as some sort of sentinel organism and reveal degradation of a medium by a given pollutant by their scarcity in the community even before the effects of the pollutant have attained large proportions.' In agreement with Landres et al. (1988), O'Brien et al. (1993) define biological indicators as 'organisms the characteristics of which reveal the presence or absence of environmental conditions that cannot be revealed in other species or in the environment as a whole.' These authors also proposed a restrictive definition of *biomonitors* as 'organisms in which one can measure changes of certain characteristics. These changes serve to evaluate the level of environmental contamination and the consequences for the state of health of other organisms or the environment as a whole.' *Sentinels* are biomonitors acting as early alarm signals. Finally, Cairns et al. (1993b) distinguish three types of indicators:

- *compliance indicators* to verify whether the objectives of maintenance and/or restoration of the quality of a site have been met;
- *diagnostic indicators* to facilitate investigation of the cause of the disturbances observed;

- *early warning indicators* to reveal the first signs of disturbances before other indicators are affected.

In the field of ecological risk evaluation, we define here the indicators of effects *in situ* as data collected on local populations living on the site (*direct* approach) or on implanted populations (animals species caged or plant species cultivated on the site-*indirect* approach). The final points correspond to various types of toxic effects:

- the response of biomarkers of effects in *individuals*;
- ecological data: effects on *populations* and variations in the structure and functioning of *communities*.

6.2. Biomarkers

6.2.1. Definitions

A biomarker is *'a variation in cellular or molecular constituents, processes, structures, or functions; this variation induced by a xenobiotic is measurable in a biological system or a sample'* (National Research Council, 1989).

Depending on what they indicate, biomarkers are classified under three principal categories: *biomarkers of exposure,* which indicate the presence of one or several pollutants in the organism, *biomarkers of effect,* which reveal the risks of toxic effects in the long term (for example, the development of cancers), and *biomarkers of individual sensitivity* (for example, markers of genetic resistance to toxins), which indicate the existence of a different sensitivity to a toxin in one part of the population. This last category is of little use in most ecological risk evaluations. The distinction between the first two categories is basically artificial. Cellular and biochemical lesions are biomarkers of exposure, because they precisely indicate exposure to a pollutant or class of pollutants, but they are also biomarkers of effect, in so far as they are the first manifestation of subsequent toxic effects.

6.2.2. Some Examples of Biomarkers

Many biological, biochemical, or immunological parameters, linked more or less directly to the mode of action of pollutants, can serve as a basis for developing and validating biomarkers (MacCarthy and Shugart, 1990; Peakall and Shugart, 1993).

Research on biomarkers in humans must use non-invasive investigations, that is, those that do not cause damage or discomfort to the individuals. This limitation is less absolute in sentinel organisms, but it becomes important in the study of species at risk that have great value, ecological or otherwise.

The *inhibition* and *induction of enzymatic systems* are often used. The *blood cholinesterases* are a good example. These enzymes are highly inhibited by organophosphorous and carbamate insecticides, even before the appearance of visible symptoms of neurotoxicity. The inhibition can be measured by a simple and quick colorimetric method, from a blood sample. This method is very widely used to monitor workers in the fertilizer and pesticide industry and workers who spray pesticides. It has also been widely used to evaluate the exposure of fishes and birds to insecticide treatments (Mineau, 1991).

Other enzymatic markers specific to exposure to metals are known, for example, δ *aminolevulinic acid dehydratase* (ALAD), which is greatly inhibited by lead.

Other enzymatic systems, such as *cytochromes P450 of liver*, are used as biomarkers of exposure as well as effect. Cytochromes P450 are a family of isozymes that metabolize most organic molecules by oxidation and thus play a fundamental role in their elimination. Some of these isozymes are induced by a large variety of industrial pollutants (PAH and organochlorine molecules such as PCB and dioxins). This induction is easily revealed by a growth in hepatic enzymatic activities such as EROD (ethoxyresorufin O-diethylase) activity. EROD activity of fish is now systematically used as an indicator of water quality in programmes for the surveillance of aquatic environments (Payne et al., 1987). It has been studied, for the same purpose, in terrestrial species such as the common rat (Fouchecourt and Riviere, 1996). Cytochromes P450 have been used as biomarkers in terrestrial species, rodents, and birds (Riviere et al., forthcoming). The activity of cytochromes P450 is more easily measured in sentinel organisms than in humans, because the usual techniques for measuring enzymatic activity necessitate the preparation of subcellular fractions of liver, a practice too invasive to be used ordinarily.

The molecule of the xenobiotic, or one of its metabolites, can be fixed firmly, by a covalent bond, to cellular macromolecules such as DNA or hemoglobin. These *adducts* can serve as biomarkers of exposure. DNA adducts particularly merit attention because, unless they are repaired by the organism, these DNA lesions are the precursors of cancer.

The usual tests of *clinical biochemistry* (blood composition, hematological parameters, plasma contents, enzymatic activities) can give useful information on the state of health of the organism and the long-term risks. The classic example of these tests is the measurement of the rate of blood cholesterol in humans. A high rate reveals a dysfunction in the organism, which indicates a higher risk of cardiovascular diseases.

6.2.3. Advantages of Biomarkers

The biomarker is designed to be a sensitive early warning system of the risk of long-term effects. It has clear advantages in decelerating diseases in humans or detecting signs of organic or functional alterations in sentinel organisms or species at risk before irreversible damage occurs to the ecosystem.

Research on biomarkers is a rapidly developing part of human epidemiology (*molecular epidemiology*). The EPA has prioritized research in biomarkers of exposure to chemical products considered to present a high risk (benzene, trichloroethylene, acrylamide, styrene, nicotine, lead) or classes of products such as PCB and dioxins. In terms of effects, biomarkers of cancer, neurotoxicity, immunotoxicity, and pulmonary toxicity, and biomarkers of reproductive and growth disorders are the primary objectives of research (Fowle and Sexton, 1992).

Biomarkers hold a considerable interest in ecotoxicology, in as much as biochemical examinations and functional explorations can be much more effective in animals than in humans. However, it is not enough to reveal their response to certain pollutants. We must also validate them, that is, establish the relation between the intensity of the effect and the intensity of the response of the biomarker and the role of factors of confusion, physio-pathological variations, and environmental variations (seasonal effects). Research on biomarkers must be given greater importance in certain plant and animal species:

- final points of evaluation;
- animal species at the top of the food chain, such as predators or fish-eating birds;
- sentinel organisms; and
- species used in ecotoxicity assays and bioassays.

6.3. Ecological Data

6.3.1. Ecological Notes

The comprehension of phenomena at levels of organization higher than those of the individual, population, community, or ecosystem requires a familiarity with some general notions of ecology. The definitions that follow are derived from the works of Barbault (1992, 1993), which constitute the best reference on the subject. For more ecotoxicological applications, the works of Ramade (1992, 1993) can be consulted.

Ecology studies the biosphere, that is, 'the layer of the planet Earth occupied and sustained by living beings', and is organized around two major axes:

- the study of biogeochemical cycles and energy flows (dynamics of ecosystems and lands);
- the study of biodemographic processes (ecology of populations and communities).

The population is 'a group of individuals of the same species occupying the same ecosystem' and the smallest object that an ecologist can consider

Table 3.5. Principal factors responsible for the organization and dynamics of populations (according to Barbault, 1992)

Probability of recolonization
Physical factors
physical structure of the medium
intervention of limiting factors (water, temperature)
variability, predictable or unpredictable, of physical factors
Biotic factors
production of medium
diversity of available resources
range of resource use of different species
interspecific competition
predation, parasitism
mutual benefit

is the population–environment system. Table 3.5 indicates the principal relations of a population interacting with its environment.

The study of populations has three levels:

- demographic analysis, descriptive study of numbers, density, and structure of populations and their temporal evolution;
- analysis of the dynamics that act on the population;
- analysis of selective pressures that affect the adaptation of populations to their environment, that is, the study of biodemographic strategies.

After the study of natural populations and biocenoses ('the set of living beings that coexist in the same space'), the study of *communities*, multi-specific sets of populations, emerged. Communities are defined as entities having a structure and functioning, or as systems of populations that are or are capable of being interconnected. The term *guild* found sometimes in the literature corresponds more precisely to 'the entire group of related species that locally exploit the same type of resources in the same way' and may be convenient to designate a combination of species of the same taxonomic group and exploiting the same responses.

The structure of communities refers to:

- *numerical* characteristics (structure of abundance, for example);
- a mode of spatial distribution; or
- a *functional* organization: relations between different populations.

6.3.2. Methods of Measuring the Structure and Functioning of Populations and Communities

Ecological studies presuppose the measurement of final points of measurement relative to:

- the structure of populations and communities; and
- their functional properties.

Barbault (1992) emphasized that numerical measurements (density, species wealth, biomass) enable us to 'compare populations *overall*, but do not directly take into account their functional structure.'

Species wealth is the number of species represented, but to consider the quantitative composition of a community, the notion of *species diversity* is substituted for it, as it takes into account the number of species and the relative abundance of each. Species diversity can be measured by indices, the most common of which are those of Simpson and Shannon, expressed as follows:

Simpson index: $I_s = 1/\Sigma p_i^2$

Shannon index: $H' = -\Sigma p_i \cdot \log 2 p_i$

where p_i is the relative abundance of species i ($p_i = n_i/N$), N is the sum of S species constituting the population or sample, and n_i is the number of the population of species i. The Simpson index varies from 1 (a single species is present) to S (all the species present have the same abundance), while the Shannon index varies from 0 to logS.

In parallel with the indices of diversity H' or I_s, the *index of equitability* E or E_s is often calculated. It relates the observed diversity to the maximum theoretical diversity:

$E = H'\log S$ or $E_s = (I_s - 1)/(S - 1)$

Equitability varies from 0 to 1; it tends towards 0 when the numbers are concentrated in a single species, and towards 1 when all the species have the same abundance.

Other indices have been used, for example, the *index of hierarchic diversity* of Osborne et al. (1980):

$HDI = H'(t) + H'(G) + H'(S)$

which incorporates three taxonomic levels: family (F), genus (G), and species (S).

Structural criteria

Species wealth and *diversity* are final points commonly used in evaluations.

Determination of *biomass* is an index of productivity (mass of animal or plant tissue constituting a population) and can be made directly, but with a precision limited by technical difficulties. Indirect measurements based on the length/size ratio or measurement of contents of chlorophyll *a* (phytoplankton) are currently used.

The presence or absence of an *indicator species* is a criterion that has often served to evaluate adverse effects on communities. This approach is derived from old studies on the systems of 'saprobes', which showed that certain species are characteristic of aquatic mediums rich in organic matter. Increase in the organic matter content satisfies the energy needs of 'tolerant' species, whereas 'sensitive' species diminish in numbers because of competition, predation, or reduction in available oxygen (Sheehan et al., 1984). This principle must be applied with caution to toxic effluents, since species that are sensitive in some circumstances are tolerant in others (Hersh and Crumpton, 1987). Indicator species have been used to characterize the toxicity of effluents (Courtemanch and Davies, 1987). The PICT method (pollution-induced community tolerance; Blanck et al., 1988) is based on the difference of sensitivity of phytoplankton communities to pollutants. Sensitive species are less abundant when pollution increases and, inversely, are found in sites less affected by the pollution. In practice, field samples of phytoplankton are taken and cultivated in the laboratory, and different toxins can then be tested on them. The samples most tolerant to toxins come from the most polluted sites.

Biotic indices (Hellawell, 1986) summarize in a single value the set of ecological information pertaining to the structure of the community. An excellent introduction to biotic indices used to determine water quality can be found in the reviews of Verneaux (1986a, b, 1995). The index of general biological quality developed by Verneaux was the subject of an AFNOR norm (indice biologique global normalisé, IBGN; AFNOR, 1992). These indices are used for the typology of aquatic mediums, but can also serve to evaluate the impact of pollution, by comparison with situations of reference, either spatial (nearby water courses, upstream sites) or temporal (progressive degradation or rehabilitation of an aquatic system), taking into account theoretical problems inherent to these approaches, such as strength of tests of hypothesis and problems of pseudo-replication (Eberhardt and Thomas, 1991).

An interesting method is the index of biological integrity (IBI) developed by Karr (1981, 1991) to determine the effects of a drop in quality of habitats of fish communities.

Suter (1993a) emphasized the advantage of methods of multivariate analysis and other methods of analysing data (in relation to indices) in identifying the drift of communities toward an abnormal state (Johnson, 1988; Smith et al., 1989), for three reasons:
- the parameters responsible for the differences are identified;
- one can choose the parameters that will be studied (in indices, these are defined beforehand);
- they lend themselves better to subsequent statistical analyses.

Another useful approach consists of evaluating the *habitat*. If it is adapted for a certain range of species, the absence of these species indicates the probable presence of disturbing agents other than the medium itself. A typical example of this system is the English RIVPACS (River Invertebrate Prediction and Classification System), which provides estimates of the presence of certain invertebrates according to 11 physico-chemical parameters (Wright et al., 1989).

An *ataxonomic approach* has been proposed to quantify aquatic populations, especially plankton. It is based on the development of techniques of electronic counting of particles and the hypothesis that physiological parameters of organisms depend first on size (here we see again the notion of dose-scaling; cf. section 4.3.5). Details of this technique are found in an article by Steinberg et al. (1994).

Functional criteria

Functional criteria are less often used than structural criteria because they require more time and expertise. Among the usable parameters are the following:

- cellular metabolism;
- rate of growth of populations;
- nutrient or matter flow (primary production, decomposition, cycles of organic and inorganic matter.

Advantages and disadvantages of ecological criteria

The advantage of these ecological approaches is still much disputed. In aquatic environments (the most researched), the usual effects of pollution on benthic invertebrate communities are reduction in biodiversity, increase in the number of opportunistic species, and tendency for growth in numbers of species of small size (Forbes and Forbes, 1994). A decrease in biodiversity can indicate pollution, but can also arise from predation or competition, or spatial heterogeneity, or may correspond to a different instance in the ecological succession (Gray, 1980), so much so that Ford (1989) concluded that indices of diversity are not reliable indicators of the effects of pollutants.

6.3.3. Polluted Sites

Final points at the population level, which can be used in a risk evaluation of pollutants, have been recommended (Suter, 1994; Table 3.6).

Techniques of collection and analysis of field data that can be used for risk evaluation of polluted soil were described in detail by Kaputska (1989). A very detailed classification and description of methods that can be used to construct a field study can be found in Eberhardt and Thomas (1991).

Table 3.6. Final points recommended at the population level (according to Suter, 1994)

Final points of evaluation

- apparent mortality
- significant reduction of productivity (relevant for commercially exploited populations)
- reduction of abundance (relevant for natural populations)
- modifications in population distribution (relevant for regional evaluations)
- extinction of species (relevant for populations in danger)
- loss of biodiversity
- drop in the 'quality' of populations

Final points of measurement

- mortality as a function of age
- fecundity
- growth

Aquatic systems

Most investigation or biological monitoring methods were developed for aquatic environments (La Point and Fairchild, 1989). Various plant and animal groups are used to calculate biotic indices: macrophytes, algae, diatoms, plankton, benthic macro-invertebrates. Micro-invertebrates (rotifers, nematodes, etc.) have an ecological interest, but they are less useful because they are difficult to identify. The quantification of fish populations has a special advantage for several reasons:

- regulatory authorities and the public easily understand the consequences that pollution can have for fish;
- fishing areas have an economic and social value (recreational and aesthetic);
- species of fish are easier to identify than those of invertebrates;
- the environmental needs of fish are known;
- fish are considered integrators of toxic effects produced at lower trophic levels.

Other characteristics of fish populations make them more difficult to study than invertebrate populations:

- they are less abundant (in principle, field samples must not tangibly alter the numbers of the population);
- they have a wider spatial distribution.

Vegetation

Vegetation is an essential component of terrestrial ecosystems. According to the species and soil characteristics, 40–85% of plant biomass is present in the soil, in direct contact with pollutants. Plants have useful characteristics that justify their inclusion in the evaluation protocols for polluted sites and soils (Kaputska, 1989):

- they respond to stresses present in the soils by modifications of photosynthesis and respiration that facilitate the absorption and metabolism of pollutants;
- they are capable of accumulating certain pollutants in their tissues;
- they transfer pollutants in the food chain;
- they stabilize soils against wind and flow erosion, thus reducing the transport of dangerous materials out of the site.

Several characteristics of vegetation are influenced by pollution:

- primary impact (direct)
  - slowing of growth;
  - variation of the structure and/or composition of the community;
- secondary impact (indirect)
  - degradation of interactions between plants and soil micro-organisms, affecting the energy flows and nutrient cycle;
  - modification of use by animals (nutrient or habitat)

Three types of methods can be used to establish the characteristics of vegetation:

- long-distance observations (by aeroplane or satellite);
- direct observations and sampling;
- evaluation of functional processes.

The first has the advantage of defining the limits of the impact and general indications of the existing vegetation. Its major disadvantage is its low limit of resolution. The second enables us to verify the general outlines defined by the observations made by air and identify species and samples to determine densities and dominant species. Finally, the evaluation of functional processes enables direct measurement of impact on vegetation and identification of probable secondary effects on animal populations and other components of the ecosystem.

Sampling techniques for vegetation are described in a review by Kaputska (1989). It is possible to measure concentrations of pollutants or contents of various components of plants on plant samples taken from the site and from control areas. Various parts of plants can be used for these purposes: roots, leaves, or stems.

Functional parameters such as the rate of photosynthesis can be measured directly in the field with portable equipment. Another technique is to measure the differences of isotopic enrichment of CO_2 (Peterson and Fry, 1987). A comprehensive review of methods of biological evaluation from plants can be found in Doust et al. (1994) and Weinstein et al. (1991).

Terrestrial invertebrates

Several invertebrate species can be found on polluted sites. The most interesting are insects and earthworms. Toxicity of a polluted site can be studied easily on earthworms caged on the site (cf. Annexure 4). The use of earthworms in biosurveillance mechanisms is described in Abdul-Rida and Bouche (1994).

Terrestrial vertebrates

According to Davis and Winstead (1980), demographic characteristics of animal populations are estimated by counting animals themselves or signs of their presence or activity (burrows, nests, faecal pellets of rodents, bird calls, tracks) according to two types of strategies: *integral counting* (of all the elements present) or *sampling* (according to transect or quadrant techniques). The techniques of capture and recapture are particularly appropriate for evaluating population numbers. The captured animals must be identified, as far as possible, by a minimum number of characteristics:

- generic and, if possible, specific identification;
- sex and age;
- sexual stage;
- weight, size, possible weight of organs.

The impact on the community level can be evaluated by different indices, species diversity, spatial distribution of individuals, and other indicators (Hair, 1980).

6.4. Ecosystem Indicators

At a high level of organization, ecosystems are characterized by:

- their physico-chemical constitution and spatial organization;
- their trophic structure;
- various parameters, of which the most important are productivity, diversity, stability, and resilience.

Productivity is linked to the type of plant formation and depends on climatic variables. *Diversity*, for an ecosystem, corresponds to species wealth, the more or less high number of species that it comprises. *Stability* signifies, strictly speaking (Barbault, 1993), the persistence of ecosystems, possibly in stable environmental conditions. *Resilience* (or homeostasis) is the capacity of an ecosystem to return to its state of equilibrium after a disturbance (cf. Cairns and Niederlehner, 1993).

The development of indicators at the ecosystem level is desirable in risk evaluation. This objective is far from having been achieved with present techniques, but we are certainly progressing towards the perfection and implementation of such indicators. For an idea of present trends, the reader can consult Cairns and MacCormick (1991), Bartell et al. (1992), Harwell et al. (1992), Burger and Gochfeld (1992), Suter (1993a), and Linthurst et al. (1995). For the reader in search of a quick review, one of the most comprehensive articles on the general problem of ecosystem risk evaluation is that of Cairns et al. (1993b), based on the analysis of a particular case, the pollution of the American Great Lakes.

4

Risk Characterization

Several definitions of *risk characterization* have already been given. The process consists of:
- estimating the extent and probability of effects;
- estimating the associated uncertainty;
- communicating the results to managers and to the public.

This last point will be addressed in chapter 5. According to the National Research Council (1994), health risk characterization comprises four elements:
- quantitative estimation of risk;
- calculation of the associated uncertainty;
- presentation of the estimate;
- communication of the entire evaluation to the people concerned.

There are two principal forms of health risk characterization:
- evaluation of the level of risk of a given exposure (measured or estimated);
- determination of environmental concentrations that should not be surpassed, depending on the acceptable risk.

The first case is typical of evaluation of a polluted site. The second corresponds to prospective evaluation of a new product or the definition of thresholds of decontamination.

These fundamental principles are applicable to ecological risk, but the aims are different. The estimation of health risk is based on different indicators, for example, the individual risk (for an individual, a 'chance' in 1,000,000 of getting cancer) or collective risk (one individual out of 1,000,000 will probably get cancer[1]), sometimes combined with a temporal

---

1: For example, the SoilRisk model (Labienic et al., 1996) is used to evaluate carcinogenic risk for humans, more precisely the carcinogenic risk of a soil contaminant. The risk is defined as 'the upper limit of the probability that an individual will develop cancer following a defined level of exposure to a potential carcinogen, during his or her entire lifetime'; this risk can be fixed at 10^{-6}, for example (EPA, 1989a).

parameter (annual risk or 'lifetime' risk). The estimation of ecological risk is based on the rate of structural and functional effects on the populations and ecosystems (for example, variation of abundance of certain populations or loss of biodiversity).

Normally, evaluation of the nature and extent of effects is accompanied by an evaluation of uncertainty. The first stage is a *qualitative* identification of uncertain elements. There are different sources of uncertainty, for example, the EPA (1989a) identified four sources of uncertainty in the evaluation of health risk of polluted sites:

- uncertainties about the physico-chemical parameters (choice of one pollutant rather than another, selection of one route of exposure rather than another, degree of confidence in certain sources of information);
- uncertainties relative to the model (does the model describe the natural phenomena adequately?);
- uncertainties about the evaluation of toxicity (for example, uncertainties associated with extrapolation methods);
- uncertainties about the risk characterization (for example, uncertainty associated with different presentations of results).

Detailed discussions of these points can be found in a document by the National Research Council (1994).

Quantification of uncertainty is much more difficult to achieve. This important element is lacking in most evaluations, either because the uncertainty is difficult or impossible to calculate, or because the evaluators have not attempted it.

For some authors, *variability* means heterogeneity of individuals with respect to one parameter, irreducible even with additional measurements, and *uncertainty* means the lack of knowledge of a poorly described phenomenon, which can be reduced by measurements (Bogen and Spear, 1987; Burmaster and Anderson, 1994). Moreover, Suter (1990b) distinguished three components of uncertainty:

- the inherently random character of prediction in environmental matter (stochasticity);
- lack of knowledge (ignorance);
- unintentional error (error).

According to the OCDE (1995), uncertainty is the result of:

- lack of knowledge;
- intrinsic variability (stochasticity);
- factors of confusion;
- lack of precision in measurements.

The most frequent method of minimizing uncertainty is the use of a *worst-case approach*, but it must be remembered that its systematic application may lead to overestimation of risk.

1. GENERAL METHODS

Three types of methods can be used to characterize ecological risk of polluted soils:

- qualitative methods;
- quotient methods (indices);
- probability methods.

Qualitative methods characterize risk in two or three categories—high and low or high, medium, and low—most often on the basis of an expert judgement.

Quotient methods are commonly used (chapter 1, section 1.5.3). In health risk, indices are frequently used, for example, the Hazard Index, obtained by dividing the daily dose of exposure by the admissible daily dose. The present practices of ecological risk are based on the same principle and on comparison of PEC and PNEC values (chapter 1, section 3.5). If the PEC/PNEC ratio is higher than 1, there is risk. If it is lower than 1, there is no risk. The differences lie essentially in the methods used to define the PEC and PNEC values and the definition of possible safety (or evaluation) factors. Safety factors used in ecological risk are reviewed in monograph no. 105 of the OCDE (1995). There is also an interesting analysis of different methods used by the EPA between 1985 and 1991 in Gaudet et al. (1994). Quotient methods apply to the danger a pollutant poses for the individual. In the case of combinations of pollutants we can add the quotients and consider that there is a risk when the sum is higher than 1, but that supposes that the pollutants responsible have been identified.

Some quotient methods are very sophisticated and require measurements at levels higher than the individual level. Gaudet et al. (1994) cite the method used by the EPA (1987a, b), which is based on three indices: Karr's index of biological integrity of fish, the index of well-being of fish, and an index representing communities of invertebrates. Three levels of risk are defined on the basis of the quotient:

- 0.1: no risk;
- 0.1 to 10: risk;
- > 10: high risk.

Levels of risk can be defined in a PEC/PNEC approach and a more refined risk formulation can thus be obtained. In the method advocated in the Netherlands, the environmental concentration of pollutant (EC) is compared to the HC5 (cf. chapter 3, section 4.4.2: DGM, 1990):

- EC/HC5 > 1: unacceptable risk;
- 0.01 < EC/HC5 < 1: a reduction in risk must be attempted;
- EC/HC5 < 0.01: negligible risk.

Most quotient methods do not calculate uncertainty, but it possible, for example by fixing higher and lower limits to PEC and PNEC, estimated or measured. Quotient methods have many advantages. They are simple, easily understood, and economical, but the use of safety factors is always tricky to justify. Their chief disadvantage is that they seem more scientific than they really are and provide single values that may be interpreted without qualification.

Probability methods are based on the analysis of:
- probability distributions (probability that an effect takes a value x for a given value of exposure);
- cumulative probability distributions (probability that an effect takes a value x higher or lower than a given value).

The results obtained by probability methods can be summarized and communicated to the manager in the form of indices (for example, the mode of distribution) or fractiles (for example, measuring the effect associated with the 9th decile). Their greatest advantage is that they allow us to easily calculate the associated uncertainty. The most remarkable example of the application of these probability methods to ecological risk are those methods developed from the work of Kooijman (1987), the objective of which is to protect 95% of the species in a community (cf. Annexure 3). At the level of ecosystems, there are a certain number of practical methods, essentially for aquatic environments. A summary description of them can be found in Suter (1993a). No matter how appealing these methods are, examples of practical applications are still rare.

2. USE OF ECO-EPIDEMIOLOGICAL DATA

Risk characterization of polluted sites is peculiar, in the sense that it involves epidemiological data. This point was brought up in chapter 1, section 2.3.2.

An excellent example of the application of eco-epidemiological methods concerns the pathological effects observed in certain populations of fish-eating birds in the American Great Lakes. The combination of these effects is known as GLEMEDS (Great Lakes embryo mortality, edema, and deformities syndrome; Gilbertson et al., 1991). This exemplary study enabled the attribution of GLEMEDS to the presence of pollutants that induce cytochromes P450 1A1.

The criteria used by Gilbertson and team to establish the causal relationship between pollutant and toxic effects are the following:
- specificity (of the effect for the cause and of the cause for the effect);
- temporality;
- strength of the association;
- constancy of the association;
- coherence: theoretical possibility, factual compatibility, biological coherence, dose–effect relation;
- predictive capacity.

These criteria are identical to criteria of causality used in epidemiological studies on human populations (Dab, 1994; another formulation of these same criteria is given by Suter (1993a).

Specificity corresponds to the appropriateness between cause and effect. Is the cause suspected of producing only this effect? Is the effect produced only by this cause?

The correlation between two phenomena does not signify that one causes the other. The possibility of coincidences is well known to epidemiologists, and they must be extremely vigilant. Suter gives the example of a comparison two animal communities, one upstream and one downstream of an effluent. Significant differences can be found in the composition of these two communities, in parallel with variations in concentrations of pollutants, without any reason to conclude that the pollutant is the causal agent responsible for the differences. In a conventional experiment, the problem of coincidences is resolved by various statistical approaches, especially replications and random attribution of treatments to experimental lots. The case of polluted sites is much more difficult, because random attribution does not exist (in a water course, the polluted site is always downstream of the unpolluted site). The absence of replication has been discussed in detail by Suter (1993a); generally, tests of hypothesis, especially tests of null hypothesis for the comparison of two sites (the null hypothesis postulates that there are no differences between the two sites), have been criticized, because of a lack of statistical validity of the tests (Eberhardt and Thomas, 1991; Forbes and Forbes, 1994). The existence of multiple causes is a particularly complex case to analyse. In certain circumstances, we may have to renounce an epidemiological approach if there are too many factors of confusion in relation to the cause suspected, which happens mostly when the extent of variation predicted is small. Suter (1993a) cites the example of effects of ozone on cultures, which are difficult to study *in situ* because the effects of variations in ozone concentrations will be of the same order of magnitude as those of other factors of variation (e.g., general effects of climate, fertilization, phytosanitary treatments).

To establish *temporality* means to determine whether the cause always precedes the effect. *Strength of the association* refers to the level of spatial coincidence between the causal agent and the effect. *Constancy of the association* signifies that the same effects will be attributed to the same causes in different studies in different regions, by different evaluation teams, etc.

Coherence denotes the possibility of explaining the causal relation in the light of existing knowledge of the following kinds:
- theoretical: are the mechanisms compatible with existing theories?
- factual: are the data compatible with existing theories?

- biological: does the causal relation have meaning in terms of avenues of exposure and cause of effects?
- statistical correlations: is the dose–response relation established unambiguously?

Predictive capacity signifies that the deductions—made from a hypothesis formulated with regard to the observed association—are subsequently verified.

Only rarely are all these criteria met, which means that very few studies demonstrate irrefutably that environmental pollutants are responsible for ecological effects.

Mathematical models of the statistical or deterministic type can be used with *in situ* indicators. For Suter, models used in retrospective evaluations are characterized by a large number of measured parameters, in relation to estimated parameters, and by the predominance of spatial models on the local or regional scale. He also notes that conventional models can be used both ways: a predictive model estimates the effects relative to a given exposure, but it can also estimate the exposure from measured effects. One can, for example, compare the significance of the estimated source and the measured source. If the levels are identical, it is a powerful argument for imputing the effects observed to this source. Another possible application of this mode of functioning of a model is to estimate environmental concentrations (for example, the environmental concentration to be attained by rehabilitation measures on the site) from concentrations without toxic effect.

◇
5
◇

Risk Formulation and Management

In risk evaluation, *formulation of the problem* is the phase in which the problem to be treated is exactly defined, that is, environmental values, mediums, or living elements to be protected are identified, as well as the particular characteristics of these values that should be measured to realize this objective. This formulation depends on the objectives of managers. In the case of soil, for example, a field may be reserved for cultivation of cereal crops, establishment of a nature reserve, or a factory. The environmental values to be protected will not necessarily be the same in all the cases, and one may either conceive a system of protection that leaves all possibilities open (concept of 'multifunctionality') or define the particular criteria for each function.

The manager has a central role. He or she must first define the final points of evaluation together with the evaluator, who is responsible for defining final points of measurement. The evaluator then assesses the level of risk and communicates the results to the manager, who must then make decisions taking into account legal and economic aspects (section 3), and also social acceptability (section 4). As practised, risk evaluation is a decision-making tool with many advantages (section 5), but it has met with some criticism (section 6).

1. ADVERSE EFFECTS AND ACCEPTABLE EFFECTS

The objective of a health risk evaluation is clear; it is the protection of human health. The manager's problem revolves around two questions:
- Is there a possibility of an adverse effect?
- If so, is the risk acceptable?

Two positions may be taken on the existence of an adverse effect:
- *Any dose is toxic. Zero risk* does not exist, except in case of zero danger, i.e., in the absence of the product. If the pollutant is present—or thought to be present—in the environment, there is always a risk, however low it may be.

- There is a *toxicity threshold* (at which there is an absence or reversibility of effects). One can therefore talk of zero risk.

The biological problem of the existence of a threshold has not been resolved and in practice, in health risk, absence of risk consists of the determination of reference doses (RfD) or *admissible* daily doses, that is, an *acceptable risk*.

Sometimes it is estimated that the ecosystem can, without serious problems, support a certain load of pollutants, the *critical load*, defined as a quantitative estimate of the exposure to one or several pollutants under which, as far as we now know, there will not be adverse effects on sensitive elements of the environment (Hunsaker et al., 1993). This notion of 'critical load' is useful and is equivalent to the notion of a threshold at the individual level, but the difficulty remains as to whether there really is a critical load and on what criteria the threshold is based (Cairns, 1992; Cairns and Niederlehner, 1993). Some authors reject even the idea of the existence of a threshold, holding that the impact of a disturbance is indelible, that ecosystems never return to their original state, and that one must not expect them to (Landis and Uy, 1995).

It is harder to define the *acceptable risk* for a plant or animal community than for humans. Health risk is essentially a risk for individuals and, as such, an adverse effect on a single individual is unacceptable, at least in theory. In ecological risk, the notion of individual risk has meaning when one fixes a population of a single species as a final point of evaluation. For example, it is relatively easy to fix as an objective the calculation of a toxic risk that a pollutant represents and to define criteria of protection for a species on the way to extinction in a given region. The notion of individual risk disappears in risk evaluation for a community. One even admits that some individuals in a population or even an entire population may be sacrificed (for example, 5% of species, cf. van Straalen and Denneman, 1989) without the community's suffering, while elimination of 5% of the human population would certainly not lead to extinction of the species, but would obviously be unacceptable.

The chief difficulty, emphasized by all the specialists, lies in the definition of what it is possible to accept in an ecosystem, which enables us to have operational criteria of quality (or value) of ecosystems. This is done, partly, by the definition of what is in store for ecosystems. Existing ecosystems depend very heavily on human activities. Most of the territory of France consists of agro-ecosystems and managed forests, which have little in common with natural territory left to itself. Long-term objectives, such as the maintenance of biodiversity[1] or the concept of sustainable development, merit a detailed discussion that the reader can find in other works.

---

[1]On biodiversity, the reader may consult the report of the Academie des Sciences (1995).

Different terms are often used to express what is desirable, such as *environmental quality* or *health of the ecosystem*. The latter expression appears seductive at first, because it gains public support without much discussion; all the more reason why we should look closely at what it implies.

2. HEALTH OF THE ECOSYSTEM

The concept of *health of the ecosystem*, developed by different authors (for example, Karr et al., 1986; Schaeffer et al., 1988; Rapport, 1989; Costanza et al., 1992; Costanza and Principe, 1995), has been criticized by Forbes and Forbes (1994) and Suter (1993b). For Suter, the health of the ecosystem is a dangerous metaphor to use, because health is a property of organisms. To use this term is to consider ecosystems as 'super-organisms', whereas their properties are different. Suter's arguments are the following:

- The limits of ecosystems are not as well defined as those of organisms.
- The structures are not identical from one ecosystem to another as they are from one individual to another of the same species.
- Their evolution is not predictable.
- They do not have mechanisms to maintain their homeostasis.

This last argument is disputable, and it would be more reasonable to say that ecosystems are not in a state of equilibrium. For Suter, the term 'health of the ecosystem' is more objectionable in that those who use it, constrained by what it implies, tend to advocate forms of management of ecosystems with strongly medical connotations ('care' of 'sick' ecosystems), which he finds unacceptable. He admits that the term 'health' has been used for entities other than the environment,[2] for example the economy ('the good health of the economy'), but in the latter case, operational concepts can be referred to, such as full employment or stable currency, that do not exist for ecosystems.

The health of the human organism is a high value for which it is possible to accept sacrifices. For example, a limb may be amputated to prevent gangrene and the death of an individual. Such a notion does not seem acceptable for ecosystems. On what basis do we decide which part of the system is better and which can be eliminated? How do we practically define a healthy ecosystem? If we admit that the health of an ecosystem is not a workable concept, we must define measurable properties of an ecosystem that can be combined to provide an index representative of good or bad health. Suter makes a detailed criticism of two systems of

---

[2]The notion of the environment itself is hard to define, inasmuch as, 'constructed from the French word environ, the term environment, of Anglo-Saxon origin, does not belong to a properly scientific vocabulary' (Berlan-Darque and Kalaora, 1989).

indices, Karr's index of biological integrity (Karr et al., 1986) and the health index of Costanza (Costanza et al., 1992), but the examination of the arguments he develops goes far beyond the scope of this work.

According to Suter, other approaches are more appropriate to ensure correct management:

- *Different criteria of measurement*, especially a better definition of final points and population indices, and methods of measurement more centred around present techniques of data analysis.
- *Different objectives*. Suter's discussion concerns essentially the definition of environmental 'quality', from the perspective of *sustainable development*.

Eventually, if the ultimate objective of ecological risk evaluation is to protect ecosystems, our knowledge remains very limited on several points:

- the exact impact of pollutants on ecosystems;
- the distinction between what is caused by pollutants and what is the impact of other stresses;
- the prediction of effects beyond the individual level;
- the definition of what is to be protected.

3. LEGAL AND ECONOMIC ASPECTS

Legal and economic aspects of the problem of polluted soils are mentioned here, because of their obvious importance. The economic aspect has been exhaustively developed in the I2C2 report (1994). The report especially addresses monetary evaluation of damages caused to the environment, economic bases of the politics of decontamination, and definition of an optimal level of decontamination. Legal aspects are treated in the same document.

4. PERCEPTION AND COMMUNICATION OF RISK

This particularly difficult point also lies beyond the limits of risk evaluation strictly speaking and merits a detailed analysis. Some indications are given below on questions concerning the communication and perception of ecological risk itself; communication and public perception of the value of evaluation strategies in resolving ecological problems are discussed in sections 5 and 6.

Forbes and Forbes (1994) discuss three decision-making groups:

- regulatory authorities;
- basic researchers (ecologists and toxicologists);
- industry.

Forbes and Forbes are two Danish ecologists, specialists in problems of aquatic ecotoxicology. In their work, they criticize the conventional predictive approach based on ecotoxicity tests, while recognizing that it is the only one possible at present. Their work also highlights the decisive role of the public, decision-makers, and the media in the definition of sensitive environmental problems and in the manner in which these problems are treated. They also propose, with some humour, to add two categories to the classical statistical errors (errors of the first and second order): errors of the third order, which occur when a problem is not highlighted that is scientifically important but politically and socially considered negligible, and errors of the fourth order, which occur when the public and decision-makers (or, according to Forbes and Forbes, certain scientists themselves, for reasons that are not always scientific) give disproportionate attention to problems that are unimportant.

To define what is significant and separate the 'true' problems from 'false' problems is also not simple (Forbes and Forbes do not give criteria of selection), but their position is interesting because it clearly expresses that environmental pollution is the concern not only of some people, but of all. The importance of recognizing the role of the public and the need for a collective effort and better communication are widely underestimated, by industry, scientists of all fields, and decision-makers. Problems of communication are very important. Even the most current practices must sometimes be called into question. For example, contrary to a well-established idea, *risk comparisons* ('the risk of smoking is much greater than the risk associated with an incinerator or an effluent') are arguments that have little efficacy in health risk (Legator and Strawn, 1993; Freundenburg and Rursch, 1994; Shrader-Frechette, 1994). Two excellent articles (Thompson, 1994; Petts, 1994) summarize public perception of waste management issues, addressing the classic NIMBY syndrome (not in my backyard) and public reactions to LULUs (locally unacceptable land uses). On problems of education and communication on the subject of environmental risk, the reader can consult Peters (1993).

These articles essentially concern health risk. Public perception of the ecological risk of pollutants has been less studied by specialists. Articles by Conn and Rich (1992), Mount (1992), Tullock and Brady (1992), Gray (1993), and Cairns (1993) can be consulted.

5. ADVANTAGE OF DEVELOPING A STRATEGY OF ECOLOGICAL RISK EVALUATION

Ecological risk evaluation of chemical products is a decision-making tool with the following objectives, summarized by Suter (1993a):
- to indicate the totality of effects;

- to justify and choose a type of action;
- to evaluate the damage done;
- to define priorities;
- to take regulatory decisions or fix standards;
- to discover the causes of an ecological disturbance (for example, the decline of an animal population).

Defining a risk evaluation strategy to tackle a problem of chemical pollution has several advantages. The first is that it neatly separates the function and responsibility of different players, especially in risk evaluation and management. The risk evaluator (or evaluation team) must provide a scientifically justified estimation, in constant dialogue with the risk manager. The risk manager represents the authority that poses the environmental problem to be resolved and ultimately takes the decision on the basis of other considerations, such as, risk–benefit analysis, psychological impact of decisions taken, or their coherence with different regulatory mechanisms.

Another advantage is that it rationally poses the environmental problem to be resolved, even if we cannot obtain the information required to resolve it completely. Unlike in other fields (insurance, for example), few data are available for risk evaluation of chemical products. The evaluation must be based on a combination of hypotheses and generalizations that modify the conclusions by a large coefficient of uncertainty and necessitate a more frequent appeal to expert judgement, despite its unscientific character. Risk evaluation stretches the limits of the available scientific data. With a very limited set of data the environmental problem cannot be resolved in a completely satisfying manner, but a risk evaluation conducted on a solid conceptual basis can indicate the fundamental information or factual knowledge needed to obtain a better evaluation.

So that risk evaluation practices—particularly for health risks—can be made more useful, Covello and Merkhofer (1993) proposed that they be based on six criteria. The first three are 'internal' criteria of analysis, pertaining to the quality of the evaluation itself, while the last three are 'external' criteria, designed to place the evaluation in the more general context of fixed objectives and the results required. These criteria are as follows:

- *Solid scientific bases.* Covello and Merkhofer remark that the validity of models used poses fewer problems than the validity of the hypotheses on which they are based and of the ways in which they are used (methods of extrapolation, for example).
- *Exhaustivity.* Risk evaluation must take into account all the practical and theoretical aspects of the problem posed.
- *Precision.* The final results must be exact and not biased.

- *Acceptability.* Risk evaluation must be understood and accepted by those it concerns, that is, by the managers who raised the problem and by the public.
- *Practicality.* The evaluation should be effectively achievable with existing means.
- *Efficacy.* The results of the risk evaluation must effectively help the manager to take a decision.

Risk evaluation procedures are currently the focus of operation of many government agencies in the United States and Europe, and of international organizations such as the OMS. It is important to understand the philosophy, principles, and present practices of risk evaluation, because these are the basis of regulatory directives, norms, and decisions that will have increasingly important repercussions on industrial and agricultural activities. Strategies of ecological risk evaluation linked to polluted soils are practised or are being developed in most industrialized countries. They are based on a common conceptual foundation but must be adapted to local situations in evaluation of products, processes, or sites.

6. EVALUATION OF CONTROVERSIAL RISKS

Risk evaluation practices are not unanimously accepted by scientists. Although detractors are still in a minority, it is important to be aware of their argument. The essence of that argument is summarized by Silbergeld (1993), basically for the evaluation of *health risk of carcinogenic products* in the United States. Silbergeld identifies three periods in the history of risk evaluation. Before 1980, the regulation of toxic chemical products was based on three principles:

- *reduction at source*, depending on technological capacities and capacity for analytic detection;
- *a risk–benefit approach*, assuming that it is possible to find a level of emission at which the benefit is greater than the risk;
- *banning of emissions*, even without the need to predict and justify the extent of possible risk (Delaney amendment).

Since the 1980s, new American legislation called into question the adequacy of these approaches and admitted the need to develop them afresh. The typical example is that of cancer. The need for prevention, in parallel with the possibility of identifying carcinogens with sufficiently simple laboratory tests, leads to several observations:

- Exposure to pollutants is a significant cause of cancer.
- Cancers of chemical origin can be prevented by identifying the products responsible.

- Evaluation of potential effects must be quantitative in order to ensure that the regulations correspond well to the criteria of risk–benefit equilibrium.

During this period, US government agencies adopted certain principles to put risk evaluation of carcinogens on a better footing.

- Risk evaluation must be able to do without inductive epidemiological data.
- There is a need for quantitative estimates of risk, that is, levels of pollution considered socially acceptable.
- The approach must be conservative.

According to Silbergeld, the scientific community remains divided about these practices. Certainly, simple acceptance of the existence of a risk can be criticized and would even be illegal because it implicitly allows that some groups will be at risk, which goes against the principle, and law, that all citizens are equal. Some advocate strict interpretations of risk evaluation, limiting it to initial phases and to animal studies, taking the position that a substance carcinogenic in an animal is dangerous under any circumstances. The basic criticism of risk evaluation methods concerns the extent of uncertainty associated with hypotheses used by default for quantifying risk at very low levels of exposure (for example, one case of cancer out of 10^6). It is generally considered that the methods used lead to very conservative results, but that also is disputed (Finkel, 1989).

However, for products that have generated the greatest controversy these past few years (Silbergeld cites the example of dichloromethane, formaldehyde, perchloroethylene, and dioxins), most critics have not made constructive propositions to develop an approach other than risk evaluation, except to demand 'better science', and no one has defined what that means. Silbergeld remarks that risk evaluation is primarily a regulatory tool, not a means to test scientific hypotheses, and that this point is not always well understood.

It is no longer obvious that risk evaluation improves the efficacy of political decisions. The chief argument invoked by Silbergeld concerns the insufficiency of data. It is very difficult to know when to stop collecting data because, for products such as those that have been named, new data appear practically every day. Risk evaluation could be seen as a diversion tactic, meant to delay regulatory action. The example she cites is that of dioxins, which were first evaluated by the EPA in the 1970s, but without any resulting regulations. A second evaluation and the determination of a dose without effect for TCDD took place in 1986, following the discovery of other sources of dioxins (incinerators) and the potential presence of these products in polluted sites. The Environmental Defense Fund instructed the EPA to regulate the sources of dioxins (chemical industry, incinerators, waste treatment plants), but the regulations were again delayed. The latest research indicated new sources of dioxins and the need for a new evaluation by the EPA. These delays prejudice the image of risk evaluations in the eyes of the public.[3] Public ignorance may be partly responsible for this but, as Silbergeld remarks, that is a poor argument. A regulatory decision must be transparent. Moreover, the public, even without being ignorant, may well be sceptical about a method that provides highly variable results, and that even specialists do not agree on.

Risk evaluation of carcinogens arose from scientific progress (which enabled the identification of carcinogens), pressures on government agencies to quantify these risks, and the public's desire to regulate chemical products on a sound basis. Silbergeld estimates that these objectives have not been reached and proposes three other approaches:

- A simple return to either regulation at source or bans. She cites regulation of DDT and PCBs and, more recently, products depleting the ozone layer, as examples of the success of this approach.
- The adoption of simpler methods to estimate risk. She gives the example of European methods of risk evaluation.
- Obligatory information to the public in case of emission and exposure to toxic products by industrial activities or other sources.

Silbergeld remarks that programmes for the voluntary reduction of toxic emissions are generally well accepted by industry and cites the example of the 33/50 programme, which consists of a voluntary reduction of 50% of dispersal of 17 priority pollutants between 1990 and 1995 with an intermediate stage in 1992 (reduction of 33%). Details about this programme can be found in Bretthauer (1993).

Silbergeld's observations (which are those of the Environmental Defense Fund) may appear exaggerated, but the simple fact that they exist forces us to take them into account. The US Academy of Sciences has moreover recognized these problems (National Research Council, 1994) and drawn up a list of criticisms about the way in which risk evaluations are conducted and their advantages in management practices. It remains to be seen whether these criticisms are useful, whether they represent a majority view, and whether they can be applied to ecological risk.

---

[3]The point brought up by Silbergeld is interesting because it highlights the *temporal* aspect. A factor to take into account in risk evaluation, and rarely pointed out, is the time required to complete it. It obviously depends on the degree of complexity foreseen at the beginning. Simple evaluations on the basis of public-domain data can be done in a month and bioassays in a few months, but field samples of animal or plant species, or follow-up studies of natural populations, require at least a year or two.

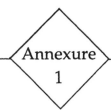

Major Mathematical Models

The following sections contain:
- a summary of models describing the occurrence and behaviour of products in soil and the models specially adapted to polluted sites;
- a summary presentation of multimedia models, especially the fugacity (or transient) models of Mackay;
- supplementary examples of algorithms for exposure of plants and animals.

It is essential to keep in mind that the choice of these examples does not imply judgement of the value or relevance of the models presented.[1]

1. MODELS OF OCCURRENCE AND BEHAVIOUR IN SOILS

These models have evolved significantly over several years (Addiscott and Wagenet, 1985; Jury and Ghrodati, 1989). Most function according to classical determinist approaches, but a stochastic approach may be an interesting means of describing phenomena in heterogeneous mediums such as soils.

Several models are presented in Table A1.1 (the essence of the information presented in this section comes from the I2C2 report, 1994). Soil is considered a homogeneous medium (BAM and PESTFADE) or a heterogeneous one. In the latter case, the pedological profile is cut into several horizontal sublayers. All the models account for water loss by evapotranspiration, except the BAM model. The modalities of water transport are described in three ways:
- water transport in stationary regime, flows being vertical and descending (BAM);

---

[1]The greatest possible care has been taken in transcribing the equations and numerical values, but the reader is advised to verify them carefully before using them.

Table A1.1. Some models used to describe behaviour of organic pollutants in soils (after I2C2, 1994)

Models	Water transport		Solute transport
	Infiltration	Evapotranspiration	
BAM	stationary water flow	none	convection
CMLS	layers of soil reservoirs —piston effect	daily	piston effect
PRZM	layers of soil reservoirs —overflow	daily	convection
CMLS	layers of soil reservoirs —overflow (mobile and immobile water)	daily (can be calculated from latitude and altitude)	convection
LEACHM	Richards equation	weekly	convection–dispersion
LEACHM	Richards equation	occasionally	convection–dispersion

- assimilation of soil layers into reservoirs, water flowing from one to the other by piston effect (CMLS) or by overflow (PRZM and VARLEACH);
- use of the Richards equation (LEACHM and PESTFADE).

Transports of solutes can be incorporated into most of these models as a function of conventional parameters (K_d and K_{oc}). Some models apply to problems of flow, sometimes taking into account interception by foliage and the half-life of pollutants on leaves.

Other models (cf. Mackay and Paterson in Suter, 1993a) include SESOIL (Seasonal Soil Transport Model; OCDE, 1989), PESTAN (Pesticide Analytical Model; Enfield et al., 1982), and the Jury model (Jury et al., 1983).

2. MULTIMEDIA MODELS

A great many multimedia models have been developed since the 1980s. They vary in complexity, ranging from measurements of equilibrium values of a non-degradable product up to static or dynamic models of occurrence of metabolites. The summary description of some existing models presented in the rest of this section is developed essentially from Mackay and Paterson (in Suter, 1993a) and Mackay (1994). The reader may also refer to specialized documents, such as the OCDE reports (1989, 1993a, c).

The *persistence model* (Roberts et al., 1981; Asher et al., 1985) developed in Canada is designed for the quick evaluation of pesticide occurrence in an aquatic environment. It is based on four compartments: water, sediments, fishes, and suspended solids (living or non-living elements). Some standard environments, a lake or a pond, are defined by default. The model takes into account photodegradation, volatilization, aqueous hydrolysis, biotransformation in fish, and microbial biodegradation in suspended solids and sediments. The results are equilibrium values and kinetic parameters.

The *GEOTOX model* (MacKone and Layton, 1986) developed in the United States calculates distribution in various mediums, reactions of degradation, and active transport and diffusion. The model was developed for the southeastern United States, but it can be adjusted for other regions. The model can be adapted to variable or constant sources of pollution and calculates the environmental concentrations at equilibrium or kinetic concentrations. The environmental concentrations can then be used to calculate the total human exposure or exposure by different avenues (e.g., inhalation, ingestion).

The *TOXSCREEN model* (Hetrick and MacDowell-Boyer, 1983) was developed by the EPA as a tool for preliminary evaluation of human exposure to organic pollutants. Soil contamination is estimated by the SESOIL model, whereas the concentrations in water are estimated by the EXAMS model. The model requires information on characteristics of the environment and emissions, data files on climatic conditions, and soil characteristics in the numerous regions included in the study.

The *EEP model* (Environmental Exposure Potentials; Klein et al., 1988) applies to continuous emissions from multiple or diffuse sources. Three types of environmental dispersions are foreseen, according to the quantity produced industrially: (1), up to 3%, the production found in the environment, (2) up to 10%, and (3) total production. The model does not calculate environmental concentrations or account for abiotic degradation. The data required are the physico-chemical properties of the products.

The *MNSEM* (Multi-phase Non-Steady State Equilibrium Model; Yoshida et al., 1987) is a simplified kinetic model, developed in Japan to predict the occurrence of products in conditions of equilibrium of a regular emission. It calculates the concentrations, distribution, persistence, and time required to return to initial level after the end of emissions. The model incorporates compartments and sub-compartments: air (raindrops), surface water (suspended solids and living organisms), soil (aqueous, gaseous, and solid phases), and sediments (interstitial water and solid phase). The processes of transfer by diffusion and active transport (rain, flow, sediment deposit, leaching of soil) are characterized by their velocity constants. The processes of degradation by oxidation, photolysis, hydrolysis, and biodegradation are taken into account in the model. The data required are the solubility, vapour pressure, standardized coefficient of adsorption (K_{OC}), and bioconcentration factor (BCF). The sources are estimated from production data.

Fugacity models represent an important class of multimedia models. They are described in greater detail in section 4.

3. MODELS ADAPTED TO POLLUTED SITES

The *AERIS model* (Senes Consultants, 1989) is a multimedia model estimating environmental concentrations and human exposure near polluted sites. It is conceived as an expert system, interactive with the user by means of questions and answers, and providing default values if needed. The model calculates equilibrium values of contaminants in mediums such as air or the phreatic layer as a function of the concentration of contaminants in the soil. It aims to provide a coherent approach to establishing directives for site rehabilitation.

The *RAPS model* (Remedial Action Priority System; Whelan et al., 1987) was developed to define priorities in identification and improvement of sites polluted by chemical and sometimes by radioactive products. Environmental concentrations are calculated in the four potentially polluted compartments: air, water soil and sediments. This model is not actually multimedia, because the models corresponding to the different mediums are independent. The advantage is that one can easily incorporate methodological innovations. The environmental concentrations help in calculating human exposure and determining the Hazard Potential Index (HPI).

To model exposure, the SEAM (EPA, 1988b) recommends the use of GEMS (Graphical Exposure Modelling System), which comprises several submodels, particularly models of occurrence of pollutants in soils, SESOIL and AT123D (Analytical Transient 1,2,3-Dimensions Model). This last model can simulate transport and occurrence of pollutants in 300 different situations, but it requires a large number of data.

4. FUGACITY MODELS

4.1. Fugacity

4.1.1. Definition

Fugacity characterizes the tendency of a product to escape from one environmental compartment (or one physical phase) to another. This physical magnitude is proportionate to the concentration and is expressed in units of pressure (Pa). The relation between fugacity (f) and concentration (C) is written as:

(1) $f = C/Z$

where Z is the fugacity constant (in mole·m^{-3}·Pa), depending on the nature of the product and the characteristics of the phase at a given temperature and pressure. A product introduced in a multiphase system tends toward equilibrium, so that the fugacities are the same in the different phases (equifugacity). The concentrations will be different, the highest Z values corresponding to the highest C values.

The values of Z can be calculated for different mediums. For the atmosphere, fugacity can be assimilated to the partial vapour pressure (P), which is itself related to the concentration by the following classic formula:

(2) $PV = nRT$

From this the following can be deduced:

(3) $C = n/V = P/RT = f \cdot Z = f/RT$

From which:

(4) $Z = 1/RT$

For other compartments, one sets off from the equation at equilibrium:

(5) $f_1 = f_2 = ... \; f_i = f_j$

where f_i and f_j are the fugacities of compartments i and j. The relation of equality between the fugacities of different compartments can be written as:

(6) $f_i = f_j = C_i/Z_i = C_j/Z_j$ or (6) $Z_i/Z_j = C_i/C_j = K_{ij}$

where K_{ij} is the distribution coefficient between the two phases. The Z value of a compartment is calculated from the partition coefficient and the Z value of another compartment:

(7) $Z_i = K_{ij} \cdot Z_j$

For example, for water:

(8) $Z_w - K_{wa} \cdot Z_a$

where K_{wa} is the distribution coefficient between air and water. From this is derived the following:

(9) $Z_w = (C_w/C_a) \cdot (1/RT)$ or   (10) $Z_w = (C_w \cdot RT/P) \cdot$

$(1/RT) = C_w/P = 1/H$

where H is the Henry constant.

Figure A1.1 and Table A1.2 give an example of calculation of the state of equilibrium of two insecticides, sulfotep and chlorfenvinphos (Calamari and Vighi, 1992).

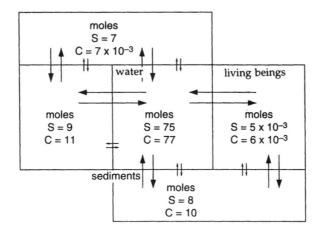

Fig. A1.1. Distribution at equilibrium of two insecticides, sulfotep and chlorfenvinphos, in a 'standard environment bracket' of Mackay. The constants needed for the calculation are found in Table A1.2 (according to Calamari and Vighi, 1992; with the permission of Scope Publications, 1997).

Table A1.2. Constants needed for the calculation of values presented in Fig. A1.1

	sulfotep	*chlorfenvinphos*
molecular weight	322	360
aqueous solubility (g/l)	0.025	0.129
partial vapour pressure	0.0226	1×10^{-4}
logK_{ow}	3.0	3.1
H (Pa · m³/mole)	0.29	2.8×10^{-4}

4.1.2. Development of Fugacity Models

Fugacity models are multimedia models, developed at the University of Toronto by Mackay and colleagues in 1980. The level I model simulates the distribution at equilibrium of a given quantity of pollutant in four environmental compartments (air, water, soil, and sediments). The level II model models and calculates the concentrations at equilibrium as a function of a continuous emission of pollutant, taking into account processes of degradation (hydrolysis, photolysis, biodegradation). The level III model is an extension of the level II model and enables accounting for a slow transfer between the different compartments and non-equilibrium conditions. The level IV model enables us to model variable emissions (Mackay, 1991; Mackay and Paterson, 1991; Mackay et al., 1992).

Several models have been derived from the level III model of Mackay, for example, the *HAZCHEM model* used in the danger evaluation approach proposed by ECETOC (1993). The *SIMPLESAL model* (OCDE, 1989), developed in the Netherlands, is a fugacity model designed to evaluate the concentrations at equilibrium or the evolution of organic models or metals. The various compartments are soil, sediments, air, water, and particles in suspension and aquatic species. It takes different phenomena into account: e.g., transport by diffusion or active transport, bioconcentration, or leaching toward the phreatic layer. The data needed are emissions levels at the source and the physico-chemical properties of the pollutants. The *Enpart model* (Environmental Partitioning Model; OCDE, 1989) was conceived as a primary tool for testing new or existing chemical products. Based on the concept of fugacity, it calculates the distribution at equilibrium or the dynamics of organic products between the different environmental compartments, persistence, and the possibilities of bioconcentration. It functions with a limited quantity of data, essentially the physico-chemical properties of the product. The results are presented not in the form of absolute values but as ratios of concentrations between the different mediums.

The advantage of fugacity models for better evaluation of ecological impact of pollutants was emphasized by Clark et al. (1988). These authors find it more useful to establish the relation between pollutant concentrations in, for example, water, air, sediments, fishes, and birds, measure pollutants in living organisms, and then deduce the environmental concentrations than, as is usually done, to deduce the concentrations in living organisms from environmental concentrations. It is more economical (concentrations in living beings being higher than in the mediums, fewer analyses are needed), and usable concentrations can be put directly to work for evaluation of environmental effects.

The use of a fugacity model, however, poses some problems. For example, the 'standard environment bracket' of Mackay (see below) incorporates essentially living aquatic elements, without accounting for plant biomass and land biomass. However, according to Calamari and Vighi (1992), the addition of a 'plant' compartment does not pose theoretical problems. All that is needed is to incorporate constants accounting for transfer towards this compartment.

4.2. An Example

Figure A1.2 shows distributions at equilibrium of a hypothetical product with a fugacity of 10^{-7} Pa (calculated in Table A1.3). In this example (Clark et al., 1988), the behaviour of products is modelled in a 'standard environment bracket' in the sense of Mackay ('a unit of world'), that is, a theoretical representation of different mediums by compartments of equal volume. The calculation of fugacity and the concept of equifugacity transience allow the instant identification of disequilibrium and future trends in the different compartments of an ecosystem. In reality, the fugacities in the different compartments are not in equilibrium, for example, in the aerial compartment because of atmospheric circulation towards the less polluted compartments. The phenomena of bioconcentration and bioaccumulation also translate into higher fugacity than the theoretical premises in the 'living beings' compartments than in the mediums.

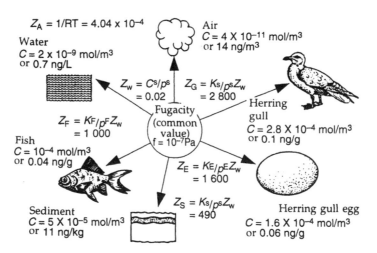

Fig. A1.2. Distribution at equilibrium of a hypothetical product of fugacity 10^{-7} Pa in a system composed of 6 compartments of 1 m^3 (data in Table A1.3; according to Clark et al., 1988; with the permission of the American Chemical Society, 1997).

5. SOME EXAMPLES OF EXPOSURE ALGORITHMS

Exposure models for different animal and plant species have already been presented in Chapter 3. The series of algorithms proposed by Hope (1995), from which some examples are described below, is selected from among existing algorithms. It is designed to provide relatively simple quantitative models for estimating the exposure of terrestrial receptors to pollutants in

Table A1.3. Distribution at equilibrium of a hypothetical compound (according to Clark et al., 1988; with the permission of the American Chemical Society)

Data needed for the calculation		
Various constants	*Partition coefficients*	*Densities*
Mol. weight = 350 g·mole^{-1}	octanol:water (K_{ow}) = 106	fish (r_t) = 1.0
Solubility (water) = 10^{-6} mole·m^{-3}	fish:water (K_{ow}) = 0.05K_{ow}[a]	seagull (r_g) = 1.0
Vapour pressure = 5×10^{-3} Pa	seagull: water (K_{ow}) = 0.14K_{ow}	seagull egg (r_e) = 1.0
R = 8.31 J·mole^{-1}·°K^{-1}	seagull egg:water (K_{ow}) = 0.08K_{ow}	sediment (r_s) = 1.5
T = 298 °K	sediment:water (K_{ow}) = 0.411 $K_{ow} \times 0.10$	
F_{oc} = 0.04		

[a]Fish is assimilated in a phase containing 5% lipids.

Calculation of concentrations in different compartments	
Air	$Za = 1/RT = 4.04 \cdot 10^{-4}$ $C = F \cdot Z = F \times 1/RT = 10^{-7} \times 1/298 = 4.04 \cdot 10^{-11}$ mole · m^{-3}
Water	$Zw = Cs/Ps = 10^{-6}/10^{-3} = 0.02$ $C = F \cdot Z = 10^{-7} \times 0.02 = 2 \cdot 10^{-9}$ mole · m^{-3}
Sediment	$Zs = K \cdot Rs \cdot Zw = 0.411 \times 106 \times 0.04 \times 1.5 \times 0.02 = 490$ $C = F \cdot Z = 490 \cdot 10^{-7} = 5 \cdot 10^{-5}$ mole · m^{-3}
Fish	$Zf = Kf \cdot Rf \cdot Zw = 0.05 \times 106 \times 0.02 = 1000$ $C = F \cdot Z = 1000 \times 10^{-7} = 10^{-4}$ mole · m^{-3}
Seagull	$Zg = Kg \cdot Rg \cdot Zw = 2800$ $C = F \cdot Z = 2800 \times 10^{-7} = 2.8 \cdot 10^{-4}$ mole · m^{-3} (0.1 mg · kg^{-1})
Seagull egg	$Ze = Ke \cdot Re \cdot Zw = 1600$ $C = F \cdot Z = 1600 \times 10^{-7} = 1.6 \cdot 10^{-4}$ mole · m^{-3} (0.06 mg · kg^{-1})

polluted sites.[2] Useful information can be found in other models, or parts of models; the HESP model (Human Exposure to Soil Pollutants, ECETOC, 1994), for example, is specifically adapted to evaluation of human risk in the case of polluted sites, but certain algorithms are useful in evaluating exposure of plants, aquatic organisms, and livestock.

5.1. Hope's Models

5.1.1. General Characteristics

The following are defined at the outset:

---

[2]Here also, we must emphasize that this selection among the existing models does not endorse their value or their relevance.

- a source and mechanism of dispersion of pollutant in the environment;
- a medium of transfer of pollutant;
- a point of exposure (receptor);
- an avenue of exposure of the receptor at the source.

The environmental concentrations (EC) are defined arbitrarily as the higher value of the interval of confidence at 95%. The exposure is estimated by application of different models (Table A1.4), but only models concerning plant and terrestrial species exposed by ingestion are presented here. Other models can be read about in Hope (1995).

Table A1.4. Hope's models (1995)

Receptor	*Avenue of exposure*	*Point of exposure*	*Model No.*
terrestrial plan	leaf deposit		1
	contact with roots	contact with soil	2 or 5
		contact with soil solution	3
		transfer to aerial parts	4
	leaf absorption (gas)		9
	leaf absorption (deposit of suspended matter)		7
terrestrial fauna	cutaneous contact		8
	inhalation (suspended matter)		9
	inhalation (vapour)		10
	ingestion	ingestion of contaminated soil ingestion of contaminated plants or prey	11
aquatic plants	direct contact		12
aquatic invertebrates	direct contact		14
	ingestion		11
aquatic vertebrates	direct contact		13
	ingestion		11

5.1.2. Terrestrial Plants

Model 1: Rainwater splash

The equation for calculating concentration of a pollutant in the aerial parts of terrestrial plants following rain splashes (as a function of the contamination of the superficial part of the soil on ~ 1 cm) is as follows:

(1) $C_{PA} = EC_{sss} \cdot K_{ps1}$

where C_{PA} is the concentration of pollutant in the aerial parts of plants ($mg \cdot kg^{-1}$), EC_{sss} is the concentration of pollutant in the soil ($mg \cdot kg^{-1}$), and K_{ps1} is the plant-soil partition coefficient. The value of K_{ps1} was estimated at $-.017$ (C_{PA} expressed in relation to fresh weight; MacKone, 1993).

Model 2: Root fixation

It is calculated by the following formula:

(2) $C_{PR} = EC_{rzs} \cdot K_{ps2}$

where C_{PR} is the concentration of pollutant in the roots (fresh weight, $mg \cdot kg^{-1}$), EC_{rzs} is the concentration of pollutant in the soil at the roots (~ 1 m, $mg \cdot kg^{-1}$), and K_{ps2} is the soil–root partition coefficient. The K_{ps2} value can be estimated by the following equation (MacKone, 1993):

(2a) $K_{ps2} = 270 \cdot K_{ow} - 0.58$

Model 3: Root fixation from soil solution

It is calculated from the following equation:

(3) $C_{PR} = EC_{sw} \cdot RCF$

where EC_{sw} is the concentration of pollutant in the surface or deep water in contact with the roots ($mg \cdot l^{-1}$), and RCF is the root bioconcentration factor estimated by the following equation (according to Briggs et al., 1982, 1983):

(3a) $RCF = 0.82 + 0.03 \cdot K_{ow}^{0.77}$

Model 4: Transfer from soil to aerial parts of plants

The equation is the following:

(4) $C_{PA} = EC_{rzs} \cdot (K_{ps3}, B_v{}^*, B_v, B_r)$

As needed, one uses K_{ps3}, plant–soil partition coefficient (soil near roots towards aerial parts), $B_v{}^*$, coefficient of bioconcentration in leaves and stems, B_v, coefficient of bioconcentration in the aerial parts, or B_r, coefficient of bioconcentration in the fruits, seeds, and tubers, with:

(4a) $\log B_v{}^* = 1.588 = 0.578 \cdot \log K_{ow}$ (on the basis of dry weight of soil and plants; Travis and Arms, 1988).

(4b) $K_{ps3} = 7.7 \cdot K_{ow-0.58}$ (on the basis of fresh weight of the plant and dry weight of soil; MacKone, 1993).

Partition coefficients to estimate the transfer of trace elements (B_v and B_r) were calculated by Baes et al. (1984).

Model 5: Transfer in the entire plant

This model, based on molecular weight, is useful when the $logK_{ow}$ is not precise. It is applied only to organic pollutants (Paterson et al., 1990).

(5) $C_{PT} = EC_{rzs} \cdot K_{ps4}$

where C_{PT} is the concentration of pollutant in the entire plant and K_{ps4} is the soil–whole plant partition coefficient, estimated according to the equation:

(5b) $logK_{ps4} = 5.943 - 2.385 \cdot logMW$

where MW is the molecular mass (g).

Model 6: Leaf fixation (gaseous phase)

Leaf fixation is important for certain pollutants (Travis and Hattemer-Frey, 1988); it can be calculated by the following formula:

(6) $C_{PA} = EC_{vap} \cdot K_{pa1}$

where EC_{vap} is the concentration of pollutant in the air ($mg \cdot m^{-3}$) and K_{pa1} is the plant–air partition coefficient ($m^3 \cdot kg^{-1}$). EC_{vap} can be estimated or measured. The most general model of distribution at equilibrium is (Riederer, 1990):

(6a) $K_{pa1} = [f_{pa} + (f_{pw} + f_{pl} \cdot K_{ow}) \cdot (RT/H)] \cdot (1/\rho_p)$

where f_{pa} is the volume of the plant in relation to the volume of air, f_{pl} is the lipid fraction, f_{pw} is the aqueous fraction, and ρ_p is the plant density ($kg \cdot m^{-3}$). According to MacKone (1993), default values can be retained as follows: $\rho_p = 1000 \ kg \cdot m^{-3}$, $f_{pa} = 0.5$, $f_{pw} = 0.4$, and $f_{pl} = 0.01$.

Model 7: Leaf fixation (deposit of suspended particles)

The formula is as follows:

(7) $C_{PA} = EC_{par} \cdot K_{pa2}$

where EC_{par} is the environmental concentration of pollutant linked to particulate matter ($mg \cdot m^{-3}$) and K_{pa2} is the plant–air partition coefficient ($mg \cdot kg^{-1}$). EC_{par} is measured at the point of exposure or estimated. The value of K_{pa2} will be 3300 for agricultural land, but less for uncultivated land, which is not subjected to the mixing caused by cultivation (MacKone, 1993; MacKone and Ryan, 1989).

5.1.3. Terrestrial Fauna

Exposure of terrestrial fauna occurs essentially by cutaneous contact or ingestion.

Model 8: Cutaneous contact

Two approaches are possible:
- contact with a volume of soil determined by the cutaneous surface and the soil depth (generally about 1 cm);
- estimation of the soil mass adhering to a given surface area of the animal.

In the first case, Dd_i, the daily dose received by the ith receptor $(mg \cdot kg^{-1} \cdot day^{-1})$, is:

(8a) $Dd_i = [SA_i \cdot CD \cdot P_c \cdot EC_{sss} \cdot CF \cdot \rho_s/W_i] \cdot \Theta_i.\Psi_i$

where

(8b) $SA_i = 12.3 \cdot (0.001 \cdot W_i)^{0.65}$

In the second case, Cd_i, the body load of pollutant, due to cutaneous contact, of the i th receptor $(mg \cdot kg^{-1})$ is:

(8c) $Cd_i = Dd_i \cdot (\partial/k_e)$

where

(8d) $Dd_i = [SA_i \cdot S_a \cdot P_c \cdot EC_{sss} \cdot CF \cdot \rho_s/W_i] \cdot \Theta_i \cdot \Psi_i$

where SA_i is the body surface of the ith receptor (cm^2), CD is the zone of contact (depth about 1 cm), P_c is the fraction of the total surface in contact with the soil, CF is a factor of conversion of units, ρ_s is the soil density $(kg \cdot l^{-1}$ or $mg \cdot cm^{-3})$, W_i is the weight of the i th receptor (kg), Θ_i is a factor of site use, Ψ_i is a seasonal factor, S_a is a factor of cutaneous adhesion $(mg \cdot cm^{-2})$, δ is a factor of cutaneous absorption of pollutant (specific characteristic), and k_e is the rate of elimination (day^{-1}).

Relation (8b) applies to all mammals and Hope (1995) indicates other algorithms to calculate the body surfaces of reptiles, birds, and amphibians. The proportion of total body surface ventrally (P_c) is estimated at 0.22 for wild rodents, but expert judgement is needed to adjust this value to other cases (e.g., birds). A value of 0.52 ± 0.9 mg of soil $\cdot cm^{-2}$ has been calculated for S_a in humans and can serve as a basis from which to calculate values for other species. The factor δ represents the fraction absorbed by the skin after contact with soil. The values existing for humans are applicable to other mammals. The rate of elimination is taken from the literature, or a default value of 1.0 is used. The factor of site use is included to take into account the duration and frequency of the presence of the receptor on the polluted site (default value 1). The seasonal factor takes into account the possibilities of migration or hibernation, which reduces the time passed on the site (default value 1).

Model 11: Ingestion

Ingestion is probably the most significant means of exposure (or at least the most often studied). For predators, exposure is caused by consumption of primary consumers, in which the degree of contamination must be evaluated. This leads to an equation of the following type:

$$\text{(11a) } Di_i = \{[(\Theta_i \cdot ECsw) + (\sum_{j=i}^{k} C_j \cdot F_j \cdot R_i) + (E_{sss} \cdot F_{sss} \cdot R_i)]/W_i\} \cdot \Theta_i \Psi_i$$

$$\text{(11b) } Ci_i = Di_i \cdot (\alpha_{ing}/k)$$

where Di_i is the dose ingested by the ith receptor ($mg \cdot kg^{-1} \cdot day^{-1}$), Ci_i is the body load of pollutant of the ith receptor ($mg \cdot kg^{-1}$), C_j is the concentration of pollutant in the jth food present in the feed of the ith receptor ($mg \cdot kg^{-1}$), F_j is the part relative to the jth food in the feeding regime of the ith receptor, R_i is the food consumption of the ith receptor (fresh weight, $kg \cdot day^{-1}$), the water consumption of the ith receptor (Θ_i; $l \cdot day^{-1}$), α_{ing} is a factor of ingestion (bioavailability), F_{sss} is the soil fraction or fraction of sediment ingested with the food, and k is the number of different foods.

If the ith receptor consumes different foods corresponding to different trophic levels, we can use a model of the trophic chain, taking into account the possibilities of biomagnification (e.g., Emlen, 1989). The following equations can be used to calculate Q_i and R_i in mammals (fresh food):

$$\text{(11c) } Q_i = 0.099 \cdot W^{0.90}$$

$$\text{(11d) } R_i = 0.054 \cdot W^{0.9451}$$

Values of the α_{ing} factor for mammals can be found in the literature (cf. Hope, 1995). For organic pollutants, the following formula can also be used:

$$\text{(11e) } \log\alpha_{ing} = -3.849 + 0.617 \cdot \log K_{ow}$$

or:

$$\text{(11f) } \log\alpha_{ing} = -2.743 + 0.542 \cdot \log K_{ow}$$

Equation (11e) provides a factor of bioaccumulation in the adipose tissue of species other than ruminants. Equation (11f) can be used for birds. A medium content of lipids of 4 to 35% is taken into account in mammals and birds to estimate the concentration in the whole organism. In the absence of usable data, the value of α_{ing} and the rates of absorption and elimination are fixed at 1.0.

5.2. The HESP Model

5.2.1. General Characteristics

The HESP model (Human Exposure to Soil Pollutants; ECETOC, 1994) is specifically adapted to the risk of terrestrial pollution:
- accidental spills;
- leaks from oil pipelines or reservoirs;
- effluent discharges;
- infiltration of chemical products into soil on storage sites;
- transport by air currents.

This model calculates the concentrations at equilibrium from a limited number of 'variable' parameters characterizing the pollutant, the soil, and the site, and 'fixed' parameters characterizing a given population in a given environment. The parameters that characterize the pollutant are:
- molecular weight;
- aqueous solubility (S_w);
- the octanol–water partition coefficient ($\log K_{ow}$);
- the adsorption coefficient (K_d);
- the partial vapour pressure (P);
- the pK_a;
- the coefficient of diffusion in air (D_a).

Various avenues of exposure are forecast. Human exposure is based on residence at the centre of the contaminated zone, at concentrations stable over time and uniform on the horizontal plan. The model evaluates the exposure of two types of human receptors, an adult and a child. The climatic conditions can be adjusted to local conditions. For the Netherlands, the following conditions have been used: a year consisting of two seasons of 6 months, a temperature of 20°C in summer and 0°C in winter, and a mean wind velocity (V_{10}) of 7.5 m s^{-1} to 10 m above soil level.

Some elements of this model are described below. They are useful for evaluations of ecological risk, for example, for herbivorous animals. The equation numbers correspond to those of the original publication (ECETOC, 1994).

5.2.2. Transfer to Plants through Roots

Metals

The formula is as follows:

(45) C_{pl} (fresh weight) = $BCF_{plant} \cdot C_s$

where C_{pl} is concentration in the plant stem (fresh weight) and C_s is concentration in the soil (including aqueous phase).

(41) $LogK_d = (3.02 - 0.85) \cdot LogBCF_{plant}$ ($r^2 = 0.68$)

This equation is based on the relation between BCF and K_d established by Baes (1982). The BCF values are taken from Sauerbeck (1988). The model also proposes equations to calculate K_d as a function of pH and the organic carbon content (Dijkshoorn et al., 1981, 1983a, 1983b).

Organic pollutants

The following equations are based on the data of Ryan et al. (1988).

(45) $C_{pl} = BCF_{stem} \cdot C_s$

(44) $BCF_{stem} = (SG/[(SG \cdot K_{oc} \cdot f_{oc}) + SN_w]) \cdot (10^{(0.9555 \cdot \log K_{ow} - 2.05)} + 0.82) \cdot (0.784 \cdot 10^{-0.4434 \, [(\log K_{ow} - 1.78)/2.44]^2})$

where SG is soil density and SN_w is water content in soil.

(47) $C_{pl} = BCF_{root} \cdot C_{pw}$

(46) $\log(BCF_{root} - 0.82) = 0.77 \log K_{ow} - 1.52$

where C_{pw} is concentration of water in pores.

5.2.3. Fixation on Plants by Atmospheric Deposit

(48) C_{pl} (dry weight) = $(f_{in}/Y_v \cdot f_{Ei}) \cdot [1 - (1 - e^{-f_{Ei} \cdot t_e})/(f_{Ei} \cdot t_e)] \cdot D_{ro} \cdot f_{rs} \cdot C_s$

where f_{in} is the intercepted initial fraction, Y_v is the plant productivity, f_{Ei} is the climatic constant, t_e is the period of growth of the vegetation, D_{ro} is the rate of external deposit, f_{rs} is the rate of soil in particular matter, and C_s is concentration in the soil. This equation is drawn from *Users Manual for TOXSCREEN* (Hetrick and McDowell-Boyer, 1984). The end value is fixed at 0.4. Tables of plant productivity values (Y_v) can be found in CSA (1986). In this model, this value is fixed at 0.28 $kg \cdot m^{-2}$. The climatic constant (f_{Ei}) varies from 0.033 to 0.05 per day, for half-lives of 21 and 15 days respectively. The duration of plant growth (t_e) is 150 to 180 days. The rate of external deposit (D_{ro}) used in the AERIS model is

230 mg $\cdot$ m^{-2} $\cdot$ day^{-1} (Reades and Gorber, 1986). It would be 43.2 mg $\cdot$ m^{-2}. day^{-1} according to Olie et al. (1983).

5.2.4. Absorption by Cattle

Quantities consumed:

(56) $TI_c = DI_c + IP_c + IV_c + VI_c + DC_{cw}$

where TI_c is total absorption of pollutant by cattle, DI_c is ingestion of soil, IP_c is inhalation of pollutant (particulate phase), IV_c is inhalation of pollutant (gaseous phase), VI_c is ingestion of contaminated plants, and DI_{cw} is ingestion of contaminated water.

(51) $DI_c = C_s \cdot AID_c \cdot (txo_c/24) \cdot f_{ac} \cdot N$

where AID_c is the quantity of soil ingested occasionally, txo_c is the time spent outdoors per day, f_{ac} is the bioavailable fraction, and N is the number of days of exposure per year.

(52) $IP_c = C_s \cdot VA_c \cdot (txo_c/24) \cdot TSP_o \cdot f_{rs} \cdot f_{rc} \cdot f_{ac} \cdot N$

where VA_c is the volume of air inhaled per day, TSP_o is the quantity of suspended particles, f_{rs} is the rate of soil in the particulate matter, and f_{rc} is the fraction retained in the lung.

(53) $IV_c = VA_c \cdot C_{oa} \cdot (txo_c/24) \cdot f_{ac} \cdot N$

where C_{oa} is the concentration in the air.

(54) $VI_c = C_{pl} \cdot Q_{pc} \cdot f_{ac}$

where Q_{pc} is the quantity of plants consumed.

(55) $DI_{cw} = [(C_t \cdot (1 - f_g - f_s) + C_{gw} \cdot f_g + C_{sw} \cdot f_s) \cdot Q_{wc}]$

where C_t is the concentration of pollutant in potable water after t days of residence, f_g is the fraction of water coming from the phreatic layer used as potable water, f_s is the surface water used as potable water, C_{gw} is the concentration of pollutant in the phreatic layer, C_{sw} is the concentration of pollutant in the surface water, and Q_{wc} is the water consumption.

These equations are identical to those used for human risk. For cattle, the following values of parameters can be used: AID_c is 0.25–0.72 kg $\cdot$ day^{-1}, VA_c is 130 m^3 $\cdot$ day^{-1}, Q_{pc} is 16.5 kg $\cdot$ day^{-1}, Q_{wc} is 55 litres $\cdot$ day^{-1}. The values of f_g, f_s, and f_{ac} are fixed respectively at 0.5, 0.5, and 0.75. The cattle are presumed to remain outdoors for 24 h/day in summer and 12 h/day in winter (McKone and Layton, 1986).

5.2.5. Concentration in Freshwater Organisms

(63) $C_m = C_{sw} \cdot BCF_m$

where C_m is concentration in the organism and C_{sw} is concentration in surface water.

(62) $\log BCF_m = C \cdot \log K_{ow} - D$

where BCF is bioconcentration factor, m is an index representing the group of aquatic organisms, and C and D are constants based on published data linking logBCF and $\log K_{ow}$ (for fish, C = 0.76 and D = 0.23; for mussel, C = 0.858 and D = 0.808; Geyer et al., 1982; McKone and Layton, 1986).

Bioassays: The Canadian Approach

The following paragraphs summarize the principles and methods used in the Canadian approach, one of the most recent in the field, to evaluate existing bioassays and propose a battery of bioassays that can be used immediately (Keddy et al., 1994, 1995).

1. OBJECTIVES OF BIOASSAYS

In the Canadian approach, bioassays have several objectives:
- to provide the data needed to propose criteria of environmental health;
- to measure the environmental health of a site;
- to indicate the need to pursue research on the site;
- to indicate the best period for rehabilitation;
- to verify the quality of the rehabilitation;
- to establish specific objectives for the site;
- to establish regulations.

All these objectives are not simultaneously covered by a single series of tests. The peculiarity of the Canadian approach is that it proposes a two-level approach based on 'preliminary' bioassays complemented by 'definitive' bioassays. Preliminary bioassays serve to prioritize the sites, demarcate a particular site, or establish specific objectives for the site. Definitive bioassays cover in greater detail the variety of biological roles and stages of development of organisms and refine the first approach. The most sensitive tests are finally used to control the efficacy of rehabilitation measures. A study of the series of tests proposed by Keddy et al. (1994) shows that the differences between the bioassays used in the preliminary evaluations and the definitive evaluations are relatively minimal, so that one might question the need to define these two stages. A point that remains to be established is the integration of data from preliminary and definitive evaluations.

2. CRITERIA OF SELECTION OF BIOASSAYS

The limits of the selection are indicated at the outset. Certain existing bioassays or ecotoxicity tests have not been included, such as tests on species that frequent the environment only temporarily (birds or bees) or tests designed to evaluate the functional capacities of soils (litter decomposition, carbon mineralization, or conversion of nitrogenous products). Three types of mediums are involved: soils, sediments, and fresh waters.

The bioassays available are numerous. To choose the most appropriate, the existing tests have been compared according to the method of Kepner and Tregoe (1965). This rather complex selection is based on the existence of two types of criteria, 'obligatory' (conditions essential to the realization of a correct test) and 'desirable' (indications that can improve the precision of the test). The desirable criteria do not all have the same weight. One attributes a score to each one, calculates the total, and converts it to a percentage (Table A2.1).

Table A2.1. 'Obligatory' and 'desirable' criteria of a bioassay (Keddy et al., 1994; with the permission of Academic Press)

Obligatory criteria
a written test protocol (e.g., methods published by organizations such as the ISO, ASTM, OCDE, or EPA)
the mention of a toxic product of reference
criteria of acceptability (i.e., criteria of good health of test organisms)

Desirable criteria
identification of one of the assay organisms of a specific level (11)
indication of a measurable final point (1)
selection of organisms (supplementary characteristics, such as size or weight) (1)
number of organisms, number of lots (1)
frequency of observations (1)
volume of experimental mediums (solids or liquids) (1)
volume of experimental container (1)
preparation of the experimental substance and mixing with the experimental medium (2)
culture and handling of colonies of organisms (1)
environmental conditions (3)
definition and manipulation of mediums (growth mediums, dilution mediums) (2)
statistical analysis (2)

Coefficient of weightage given to each criterion (total score: 17)

2.1. Evaluation: Step 1

The 'exploitable' bioassays are first identified according to published data or reports from official organizations, experts, researchers, and others. They are classified into two groups according to whether they satisfy the first obligatory criterion or not (existence of a published assay protocol).

If the first criterion is fulfilled, one then verifies the existence of two other obligatory criteria. If they are met, the bioassays are considered 'potentially exploitable'. Otherwise, they are considered 'potential prototypes'. In either case, they are subjected to a detailed evaluation on the basis of desirable criteria. According as the grade obtained is higher or lower than 88%, the 'potentially exploitable' are classified as either 'exploitable' or 'prototypes of priority 1'. The 'potential prototypes' are classified as 'prototypes of priority 2' (grade 100%), 'prototypes of priority 3' (grade between 88% and 100%), or 'in development of priority 4' (grade less than 88%). Keddy et al. (1994) propose research to improve the prototype tests and make them meet a better standard, but this point will not be discussed here.

2.2. Evaluation: Step 2

For the definitive selection of tests that can constitute a battery, additional information is collected on tests having scores higher than 88%:
- examination of the trophic level;
- sensitivity;
- reproducibility;
- field validation;
- ecological relevance.

This additional information is used not in the initial evaluation of tests, but to verify the applicability of bioassays according to the field conditions and objectives.

2.3 Series of Tests

The combination of processes leads to the recommendation of an 'exploitable series' of bioassays for each medium, complemented by an 'augmented series'. These two series are proposed for each type of evaluation, preliminary and definitive.

The process of selection is highly dependent on a hypothesis, not conclusively proved, that can be summarized thus: the best methods are those that are most often used. The merit of this criterion is debatable. Is the method most often employed because it is good or is it good because it is the most often employed? Moreover, the choice of bioassays is influenced by American studies. Among the translators of the Canadian report is J.C. Greene, who was credited as expert in the EPA report of 1989 (Warren-Hicks et al., 1989). This document represents a very interesting advance, but a thorough examination will be needed to adapt it to European conditions.

3. SOIL BIOASSAYS (Figs. A2.1, A2.2)

In the initial phase, 47 species belonging to 9 large groups of organisms were identified. Written test protocols exist for algae (1 species), vascular plants (24 species), earthworms (4 species), and collembola (1 species).

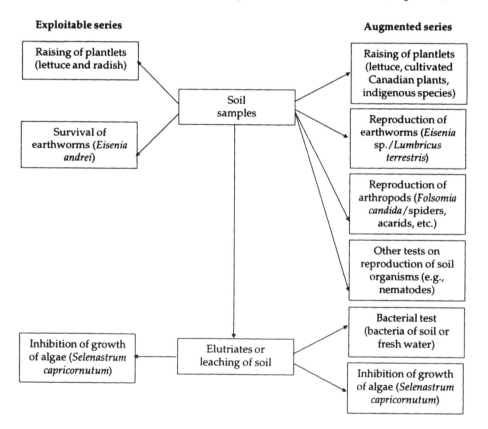

Fig. A2.1. Bioassays recommended for preliminary evaluation of soil quality (according to Keddy et al., 1995; with the permission of Academic Press, 1997).

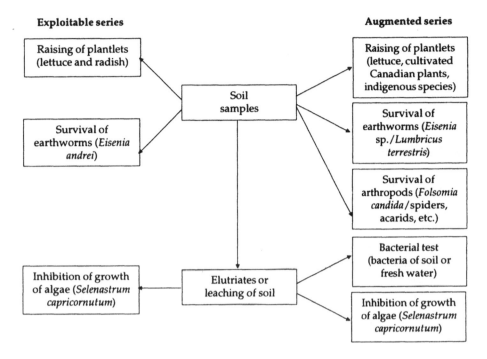

Fig. A2.2. Bioassays recommended for definitive evaluation of soil quality (according to Keddy et al., 1995; with the permission of Academic Press, 1997).

3.1. Preliminary Evaluation: Exploitable Series

The exploitable series (all the obligatory criteria and 88% of the desirable criteria are satisfied) comprise three bioassays: the raising of plantlets, the survival of earthworms, and the inhibition of algal growth.

The assay on *inhibition of growth of algae (Selenastrum capricornutum)* was integrated with the bioassays on soils to evaluate the toxicity of lixiviates and elutriates. Some technical modifications are proposed.

The assay on *inhibition of plant growth* was retained according to the protocol proposed by Greene et al. (1988b). It was considered more useful than the bioassay on root elongation, because the exposure occurs through soil rather than from aqueous extracts, which are less representative of the totality of pollutants present. The two most suitable species are lettuce and radish, but other species are possible.

The assay on *survival of earthworms* uses compost worms *E. andrei*. Two protocols are available: one proposed by the EPA (Greene et al., 1988b) and one by the ISO (1991a). These bioassays run for two weeks. In the

first, the soil from the site is diluted with an artificial soil and in the second, liquid substances are incorporated into an artificial soil. Ultimately, the EPA test was used for four reasons:
- it is standardized;
- it is used world-wide;
- it has good reproducibility;
- it has been specially designed to evaluate contaminated sites.

3.2. Preliminary Evaluation: Augmented Series

Keddy et al. (1994) noted that all the trophic levels are not represented in the exploitable series and proposed to add certain groups of organisms. The complete series comprises five tests:
- A bioassay of plant growth inhibition, incorporating an indigenous species such as soya or maize. An economically important plant such as wheat is useful, but it often has a lower sensitivity to pollutants.
- A bioassay on earthworms. *Lumbricus terrestris* is a good candidate, since it is a local species. Environnement Canada developed studies to compare the sensitivity of *E. foetida* and *L. terrestris*. If the latter is more sensitive, it will probably be necessary, if possible, to develop a protocol to use it as a test organism.
- A bioassay on arthropods, which could be based on collembola, diplopods, isopods, or other species.
- A test on bacteria, such as Microtox^{TM} or Chromotest^{TM}.
- The test for algal growth inhibition described earlier.

3.3. Definitive Evaluation: Exploitable Series

The bioassays defined for the definitive evaluation are the same as those used earlier.

3.4. Definitive Evaluation: Augmented Series

The tests used are slightly different from those used for the augmented series designed for preliminary evaluation. The modifications pertain to the bioassay on raising of plantlets that can be done on indigenous species, cultivated or otherwise. The bioassay on reproduction of earthworms (ISO, 1991b) replaces the 14-day test recommended in the exploitable series. The test on collembola (reproduction of *Folsomia candida*) is a useful candidate. Other terrestrial organisms can be used. A test on nematodes is being studied in the Netherlands.

3.5. Definition of Solid and Liquid Mediums

The success of these bioassays requires that very precise conditions be defined for preparation, storage, and collection of soil samples, as well as for preparation of elutriates and lixiviates. Three types of soils can be used: standardized artificial soil (international norm), artificial soil adapted to local needs (Canadian, in this instance), or natural soil. The artificial soil serves as negative control and possibly for dilution of the soil studied. It is used for assays on terricolous animals and on plants.

4. BIOASSAYS OF SEDIMENTS (Figs. A2.3, A2.4)

In the initial phase, 19 usable species belonging to 8 large groups of organisms were identified. Written test protocols exist for the algae

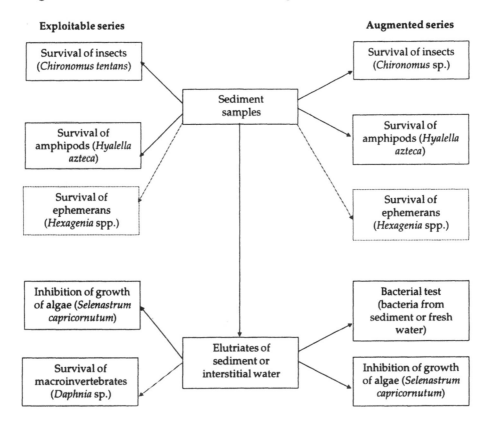

Fig. A2.3. Bioassays recommended for preliminary evaluation of sediment quality (according to Keddy et al., 1995; with the permission of Academic Press, 1997).

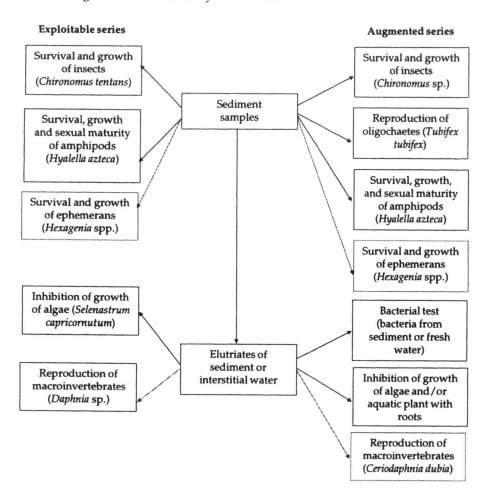

Fig. A2.4. Bioassays recommended for definitive evaluation of sediment quality (according to Keddy et al., 1995; with the permission of Academic Press, 1997).

(1 species), vascular plants (24 species), amphipods (1 species), oligochaetes (2 species), ephemerans (2 species), and dipterans (2 species).

4.1. Preliminary Evaluation: Exploitable Series

The authors emphasize the need to include organisms living in the sediments, rather than conducting assays only on aquatic organisms in the presence of elutriates or interstitial water taken from sediments. They refer to the works of Ankley et al. (1991), who demonstrated that, in the latter case, the degree of toxicity for benthic organisms may be underestimated.

The following bioassays are recommended for the exploitable series[1]:
- survival of dipterans (*Chironomus tentans*);
- survival of amphipods (*Hyalella azteca*);
- survival of ephemerans (*Hexagenia* sp.); (This test is shown in a broken-line box in Figs. A2.3 and A2.4 because it is recommended only for sites on which this species is a typical element of the benthic biomass.)
- inhibition of growth of algae (*S. capricornutum*);
- 48 h bioassay on daphnia (if data are not found on the reference products for the bioassay on *H. azteca*).

4.2. Preliminary Evaluation: Augmented Series

The augmented series has some modifications. The bioassays on dipterans (*C. tentans*) must integrate other species, such as *C. riparius* (if it proves to be more sensitive).

4.3. Definitive Evaluation: Exploitable Series

The bioassays designed for definitive evaluation are roughly the same, except for the one on amphipods, which is developed to include other final points (growth and sexual maturity) and the 48 h daphnia bioassay, which is replaced by a 7-day bioassay.

4.4. Definitive Evaluation: Augmented Series

In relation to the augmented series retained for the preliminary evaluation, a test of reproduction of oligochaetes (*Tubifex tubifex*; ASTM, preliminary test) is provided for. The assay on ephemerans (*Hexagenia* sp.) must be expanded to include growth as a final point. The assay on amphipods must be expanded to include growth and sexual maturity. The assay on daphnia is shortened. Finally, an assay for aquatic plants with roots can complement or be substituted for bioassay of algal growth inhibition.

4.5. Definition of Liquid and Solid Mediums

As with soils, sediments must be collected, stored, and prepared under standardized conditions.

---

[1]For sediments, the selection threshold of 88% is reduced to 80% because very few tests reach the 88% threshold.

5. BIOASSAYS FOR FRESH WATERS (Figs. A2.5, A2.6)

The bioassays used to evaluate the quality of fresh waters are much more numerous than those for soils and sediments: 119 usable species belonging to 212 large groups of organisms have been identified in the initial phase. Written test protocols exist for bacteria (5 species), algae (26 species), invertebrates (30 species), amphibians (2 genera), vascular plants (1 species), and fishes (25 species).

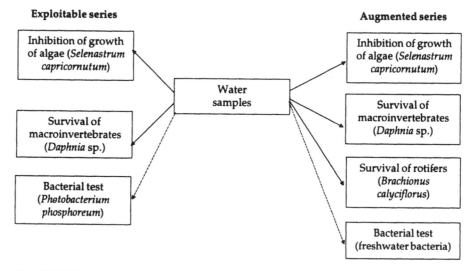

Fig. A2.5. Bioassays recommended for preliminary evaluation of the quality of fresh waters (according to Keddy et al., 1995; with the permission of Academic Press, 1997).

5.1. Preliminary Evaluation: Exploitable Series

Three bioassays have been retained:
- inhibition of growth of algae (*S. capricornutum*);
- a 48 h daphnia test (*Daphnia magna* or *D. pulex*);
- a test on *P. phosphoreum*.

In this series, a 7-day bioassay of fathead minnow (*Pimephales promelas*) is planned which would increase the number of trophic levels, but the daphnia test is considered a correct substitute, since correlation between the LC$_{50}$ values is good.

5.2. Preliminary Evaluation: Augmented Series

The augmented series incorporates an assay on rotifers (24 h test on *Brachionus calyciflorus*). The bacteria test can be based on *E. coli* or

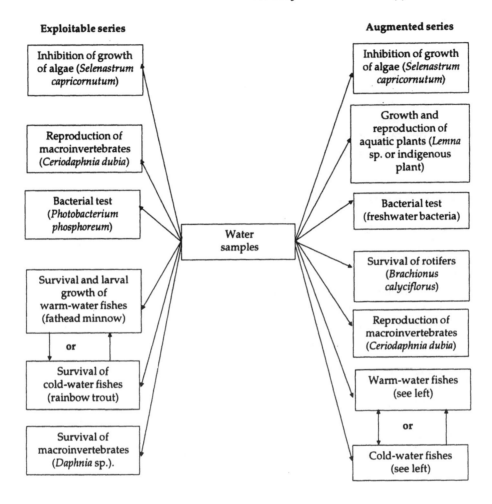

Fig. A2.6. Bioassays recommended for definitive evaluation of the quality of fresh waters (according to Keddy et al., 1995; with the permission of Academic Press, 1997).

Pseudomonas putida, but *P. phosphoreum* would be used, since studies on the first two species would not be complete.

5.3. Definitive Evaluation: Exploitable Series

The exploitable series is enriched by a bioassay on *C. dubia* (7 days) and a test on cold-water fish (rainbow trout) and warm-water fish (larva of fathead minnow).

5.4. Definitive Evaluation: Augmented Series

Compared to the augmented series proposed in the preliminary evaluation, it is proposed to add fish bioassays of the exploitable series, replace the test on *Daphnia* sp. by one on *Ceriodaphnia* (7 days), and add a test on growth and reproduction of aquatic plants (*Lemna gibba*). This last bioassay will be considered exploitable when it is more complete, but eventually it will be desirable to have a standardized bioassay for *L. minor*, a plant indigenous to Canada.

6. RESEARCH PRIORITIES

Improvement in the series of bioassays proposed above would require a research programme. The authors have identified 20 major points that should be the object of priority research. These points are grouped here under five objectives:
- to complement certain assays: e.g., add toxic products of reference, evaluate their relative sensitivity, specify the species.
- to improve methods of collecting, preparing, and storing mediums: e.g., choosing typical soils.
- to perfect assays for certain types of organisms and mediums: e.g., aquatic plants with roots.
- to re-evaluate assays on bacteria species;
- to prepare a manual on statistical methods.

According to Keddy et al. (1994, 1995), this last point is often the weakest. Active effort must be taken to ensure that those who conduct the assays, who are not specialists, provide information that is clear and usable. The authors strongly recommend that appropriate software be developed so that the statistical methods proposed in the experimental protocols are applied in the best conditions. The report presented by Keddy et al. (1994) has an annexure containing a very interesting statistical analysis of exploitable tests.

Annexure
3

The Van Straalen and Denneman Model (1989)

The model of van Straalen and Denneman is among the methods designed to extrapolate information obtained from ecotoxicity tests on some species to the entire community. It is particularly interesting because the authors have used terrestrial ecosystems as examples.

The van Straalen and Denneman model (1989), derived from the Kooijman model (1987), is based on the following hypotheses and principles:

- the community is formed from a group of independent populations;
- the distribution of sensitivities is symmetrical on a logarithmic scale.

The sensitivities are values of final points such as NOEC, LD_{50}, and MATC. The objective of the Kooijman method is either to calculate a safety threshold (an environmental concentration) such that *the most sensitive species of the community* is protected (the sensitivity values are higher than this dose), or inversely to calculate the percentage of the population that will be protected at a given environmental concentration. The method consists of establishing the distribution curve of sensitivities from sensitivity values (NOEC or others) measured directly or extrapolated from some species of the community.

This approach has been criticized by van Straalen and Denneman and others, who remark that it is too conservative. The probability of finding one species more sensitive increases with the number of species in the data sample, that is, the more one adds to the database, the more one risks obtaining low safety thresholds.

For van Straalen and Denneman, to protect a community it is not necessary to protect the most sensitive element. Mild effects can be compensated for by the capacity of the system to recover, and the chief improvement made on the Kooijman model by van Straalen and Denneman is to render the security factor calculated less sensitive to extreme values of the frequency distribution curve and independent of the number of species to be protected in the community.

The van Straalen and Denneman model uses NOEC rather than LD_{50} values, the latter being considered less representative of natural situations. Several operations are needed to develop the model:

- standardize the data as a function of soil type;
- apply a safety factor designed to take into account different sensitivities of different species;
- apply a supplementary extrapolation factor designed to account for differences between the field and the laboratory.

1. STANDARDIZATION OF DATA

The toxic properties of products differ according to the physico-chemical characteristics of soils. In the example cited by van Straalen and Denneman, the calculations are based on the existence of reference levels considered acceptable in soils and of equations designed to calculate acceptable concentrations in other types of soil as a function of two chief variables, the clay content and the organic matter content. For example, the acceptable value of cadmium in soils is:

(1) $R = 0.4 + 0.007(L + 3H)$

where R is acceptable cadmium content ($\mu g/g$), L is clay content (%), and H is organic matter content (%). The standard soil is defined with $H = 10$ and $L = 25$, which gives, on application of equation (1), a reference value of 0.8 $\mu g/g$ of cadmium.

The standardized NOEC values ($\hat{NOEC}$) for the standard soil are calculated by the following formula:

(2) $\hat{NOEC} = NOEC(L,H) \times R(25,10)/R(L,H)$

where NOEC(L,H) is NOEC value for the experimental values of L and H, R(25,10) is reference value for the standard soil, and R(L,H) is reference value in the experimental conditions calculated by equation (1).

Van Straalen and Denneman remark that other parameters can be taken into account to improve the model (e.g., soil pH).

2. INTRASPECIFIC DIFFERENCES IN SENSITIVITY

The Kooijman model defines a concentration dangerous for the most sensitive species (HCS), while the van Straalen and Denneman model defines an HCp, that is, a concentration dangerous for p% of species in the community. The authors presume that the distribution of NOEC values follows a log-logistic distribution: when a species is chosen at random, the probability that its LogNOEC is between x_1 and x_2 is:

$$\int_{x_1}^{x_2} f(x) \cdot dx$$

where $f(x)$ is determined as follows:

$$(3) \quad f(x) = \frac{e^{\left(\frac{\mu - x}{\beta}\right)}}{\beta\left[1 + e^{\left(\frac{\mu - x}{\beta}\right)}\right]^2}$$

where μ is the mean and β is the range of the distribution.

The concentration dangerous to $p\%$ of species (HCp) is the value of x such as the probability of choosing a species with a NOEC lower than the HCp is equal to δ_1, a small number arbitrarily defined as equal to $p/100$ (e.g., 0.05), which yields the following formula:

$$\delta_1 = \int_{-\infty}^{\text{Log } HC_p} f(x) \cdot dx$$

where $f(x)$ is defined by equation (3). After integration:

$$\delta_1 = \frac{1}{1 + e^{\frac{(\mu - LogHC_p)}{\beta}}}$$

and rearrangements:

$$(4) \quad HC_p = e^{\left[\mu - \beta \text{ Log}\left(\frac{1 - \delta_1}{\delta_1}\right)\right]}$$

The parameters μ and β are estimated from a series of NOEC values deduced from chronic toxicity assays. One supposes that m species have been tested and that they represent a random sample of the community. The mean of LogNOEC values is written as x_m and the standard deviation of these values is s_m. The values of μ and β can be estimated from x_m and s_m by:

$$\hat{\mu} = x_m \text{ and } \hat{\beta} = (s_m \sqrt{3})/\pi$$

The estimate of HCp is as follows:

$$(5) \quad \hat{HC}_p = e^{\left[x_m - \frac{s_m \sqrt{3}}{\pi} \text{ Log}\left(\frac{1 - \delta_1}{\delta_1}\right)\right]}$$

The error on the estimation of HCp is fairly large: the probability that true HCp will be still smaller than the estimated HĈp is 50%. There is thus a non-negligible probability that more than p% of the species of the community will be affected when HCp is estimated by equation (5).

One of the sources of error in determining HĈp is limitation of the number of species to μ, which is the cause of an error in μ as well as β. Kooijman (1987) shows that only the error on β is important for estimating HCp and, to avoid it, a second logistic distribution can be introduced, with the same mean as the first, but with a greater standard deviation (Fig. A3.1). If the standard deviation of this distribution is equal to βd_m, then the probability of overestimating HCp can be fixed at a small number, δ_2. The factor d_m is estimated by computer simulation, supposing that μ and β are random variables that follow a logistic distribution: d_m depends on m and d_2 is provided in a table (Kooijman, 1987); d_m diminishes when m increases and $d_m \to \pi/\sqrt{3}$ when $m \to \infty$

The parameter β' of the second distribution can be estimated by:

$$\beta' = \frac{s'_m \sqrt{3}}{\pi} = \frac{\hat{\beta} d_m \sqrt{3}}{\pi} = \frac{3 s_m d_m}{\pi^2}$$

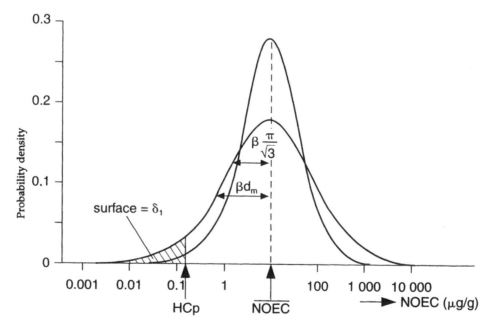

Fig. A3.1. Theoretical representation of density distributions of probability of NOECs and different parameters used in the equation of van Straalen and Denneman (van Straalen and Denneman, 1989; with the permission of Academic Press, 1997).

Equation (4) thus becomes:

$$(6) \quad \hat{HC}_p = e^{\left[x_m - \frac{3s_m d_m}{\pi^2} \, \text{Log} \left(\frac{1-\delta_1}{\delta_1}\right)\right]}$$

which can be written as:

$$(7) \quad \hat{HC}_p = \frac{e^{(x_m)}}{T} = \frac{\overline{NOEC}}{T}$$

where NOEC is the geometric mean of NOEC and T is a safety factor calculated by the following formula:

$$(8) \quad T = e^{\left[\frac{3s_m d_m}{\pi^2} \, \text{Log} \left(\frac{1-\delta_1}{\delta_1}\right)\right]}$$

To summarize, from the theoretical bases developed by Kooijman (1987), equation (8) allows calculation of a safety factor protecting $(100 - p)\%$ of species in a community. This factor is applied to the geometric mean of NOEC values (equation 7), and it is dependent on the standard deviation of NOEC values (s_m), the percentage of species not protected (δ_1), the number of species tested[1] (m), and the probability of overestimating HCp (δ_2).

Conversely, the model can be used to calculate the risk associated with a certain environmental concentration of pollutant, c. If we put in value c at the left side of equation (6) and rewrite the equation, we obtain:

$$(9) \quad q = 100 \left(1 - \left[1 + e^{\left\{\frac{\pi^2 \cdot (x_m - \text{Log } C)}{3s_m d_m}\right\}}\right]^{-1}\right)$$

where q is the percentage of species protected when the environmental concentration equals c.

3. NUMERICAL EXAMPLE

The example given by van Straalen and Denneman is designed to test the preceding formulas. It is based on the NOEC values for 7 species of terrestrial invertebrates (Tables A3.1, A3.2). The level of protection has been chosen as equal to 95% ($\delta_1 = 0.05$), which allows the calculation of the corresponding HCp value (HC5). δ_1 is chosen as equal to δ_2.

1: It is recommended that at least five different species be used.

Table A3.1. Example of calculation of a concentration that protects 95% of a population of soil invertebrates (according to van Straalen and Denneman, 1989; with the permission of Academic Press, 1997)

Parameter	Symbol	Value	Mode of calculation
Number of species tested	m	7	according to original data
Mean of LogNOÊC	x_m	2.236	according to original data
Geometric mean (mg/g)	NOEC	9.4	$e^{(xm)}$
Standard deviation of LogNOÊC	s_m	1.618	according to original data
Percentage of species not protected by HCp	δ_1	0.05	
Probability of overestimation of HCp	δ_2	0.05	
Factor dependent on m and δ_2	d_m	2.82	according to table of Kooijman (1987)
Safety factor applied to the mean of NOÊC	T	59	equation (8)
Concentration dangerous to 5% of species (μg/g)	HC5	0.16	equation (7)
Percentage of species protected with the reference value proposed for cadmium (0.8 μg/g)	q	85%	equation (9)

Table A3.2. Original data (simplified)

Species	NOÊC (mg/g)
1	15.4
2	13.5
3	13.8
4	3.63
5	3.33
6	0.97
7	18.7

Annexure 4

A Case Study

The case described below concerns a site of limited size (some 20 ha) touching a humid zone and a forest and characterized by highly varied pollution (Menzie et al., 1992). Other examples of polluted sites, aquatic and terrestrial, are available in the literature, in Suter and Loar (1992), Gray (1993), Gray and Becker (1993), Harris et al. (1994), and Pastorok et al. (1994). The most complete, too complete, in fact, to be analysed here, is that reported by Pascoe and Dal Soglio (1994a, b, c) in a series of articles.

1. THE 'BAIRD AND McGUIRE' SITE

The polluted site, located in Massachusetts (USA), about 30 km south of Boston, comprises 22 ha of woods and forests along the Cochato river. The pollution is general, resulting from 16 years of pesticide packaging activity on the site, till the 1970s. Preliminary studies revealed the presence of 102 priority pollutants of the EPA in the surface waters, the phreatic layer, the soil, and the sediments, chiefly DDT and its metabolites[1] (henceforth called 'DDT residues' or DDTR), chlordane, PAHs, and arsenic. In fact, the essence of the analysis concerned evaluation of the effects of DDTR. For the purpose of the experiment, two secondary sites were defined in the polluted zone (highly polluted and moderately polluted), and a reference site at the border of the polluted zone (with no pollution). For the capture of birds, a supplementary reference site was chosen outside the polluted zone. The soils were characterized by their total organic carbon content, their granulometry, and their pH according to the standardized methods of the EPA.

---

[1]The pollution was considerable on the highly polluted site, on the average 140 mg · kg^{-1} of DDTR, surpassing 800 mg · kg^{-1} in one sample. In the moderately polluted site, the average was 7 mg · kg^{-1} (maximum 43 mg · kg^{-1}).

2. OBJECTIVES OF THE STUDY

This study is particularly demonstrative of the procedure to be applied to contaminated sites, because its objective was to compare four types of methods[2]—prospective methods based on ecotoxicity assays, bioassays, sentinel organisms, and biological surveillance of the environment—in order to answer two questions:

- What is the capacity of pollutants to bioaccumulate in the terrestrial fauna and to biomagnify in the trophic chain?
- What are the risks resulting from exposure of living organisms to these products?

The methods were evaluated on the basis of three criteria:

- Is the method *sensitive* in detecting changes in environmental conditions, changes significant enough to be felt at the level of populations and the ecosystem?
- Is the method *practical*, in terms of means and cost, for use on other sites?
- What is the *ecological relevance* of the method, in other words, can potential effects on populations and ecosystems be interpreted from the results?

3. METHODS USED

3.1. The Predictive Model Approach

Models were chosen that were as close as possible to those recognized by the EPA in the *Superfund Exposure Assessment Manual* (EPA, 1988) in accordance with the *Supplemental Risk Assessment Guidance for the Superfund Program* (EPA, 1989b) and were applied to different animal species.

Earthworms

A bioaccumulation model based on that of Markwell et al. (1989) was developed for earthworms. This model estimates the distribution of pollutants in the soil, soil water, and worm tissues according to the following equation:

$$C_S = x \cdot (K_{ow})^m \cdot f_{oc} \cdot C_{WI}$$

where C_S is the concentration of the product in the soil (ng $\cdot$ g^{-1}), x is a constant of proportionality, K_{ow} is the octanol–water partition coefficient,

---

[2]A comparative analysis of this type, for the aquatic environment, can be found in Pontasch et al. (1989).

m is a constant, f_{oc} is the organic carbon content of soil, and C_{WI} is the concentration of pollutant in the interstitial water (ng · ml^{-1}). According to Markwell et al. (1989), the empirical results indicate values of 0.66 for x and 1.029 for m.

The relation between the concentrations in the living organisms and interstitial water is given by the following equation:

$$C_B = C_{WI} \cdot Y_L \cdot (K_{ow})^p$$

where C_B is the concentration of pollutant in the organism, Y_L is the lipid content of the organism, and p is a constant (which will equal 1 in most scenarios, according to Markwell et al., 1989).

In the present example, the values of p and m are fixed at 1 and that of Y_L is equal to 2% (Stafford and Tacon, 1988). The factor of bioaccumulation is expressed finally by the following equation:

$$BCF = Y_L \cdot (0.66 f_{oc})^{-1}$$

Note that the value of K_{ow} has disappeared from the final expression of bioaccumulation.

Birds and mammals

The bioaccumulation model used is a simple model based on the degree of pollution in the animal feed:

$$C_b = (BCF) \cdot (C_f) \cdot (F_{frac})$$

where C_b is the concentration of pollutant in the entire animal (fresh weight), BCF is bioconcentration factor of the feed (dry weight) in the animal (fresh weight), C_f is the concentration of contaminant in the feed (dry weight), and F_{frac} is the fraction of feed made up of contaminated feed.

The more sophisticated models, such as that developed by Clark et al. (1987), are not used here, as the authors estimate that they require too many data, which are difficult to obtain.

Published data were used to calculate the bioconcentration factors for pesticides in birds (Mahoney, 1974) and wild rodents (Forsyth and Peterle, 1984; cf section 4.4). Some values available in the literature (especially Garten and Trabalka, 1983) were not used, being unsuitable for the resident species.

To evaluate the dangers to birds in the present study, the authors simply defined a hazard index calculated by a quotient method according to the following formula:

$$HI = D_{feed}/D_{final\ point}$$

where D_{feed} is the dose of pollutant in the feed (concentration of pollutants in the different elements of the feed ration) and $D_{final\ point}$ is the dose in the feed corresponding to the LOAEL or the NOAEL. A ratio higher than 1 indicates a potential risk, but the authors emphasize that this index is a simple reference point that does not measure effects at the population level.

3.2. Bioassay Approach

Bioassays were used to evaluate the toxicity of soils by measuring the concentration of pollutants in the earthworms. Soil samples were collected at random in the different sites and homogenized. The animals were exposed to pure or diluted samples. The tests were done in two different laboratories with two earthworm species (*Eisenia foetida* and *Lumbricus terrestris*) according to EPA protocols (Greene et al., 1988b).

3.3. Sentinel Animal Approach

The objective was the same as before, but with direct exposure of earthworms to polluted soil. Plastic containers with polluted soil and locally raised worms (*L. terrestris*) were placed at locations from which the soil was collected. These containers were perforated to allow the passage of gas and water while keeping the worms inside. The contents of the containers were observed 1 and 7 days after the beginning of the assay. A system of notation based on the visual appearance and mortality rate of worms was used for the results and then treated with an algorithm developed for the purpose, to provide a spatial distribution of the most toxic zones from 67 sampling points.

3.4. Biological Surveillance Approach

The biological surveillance involved levels of pollutants (essentially pesticides) in plants and animals:
- Fruits, leaves, and roots from local vegetation.
- Samples of terrestrial invertebrates collected at the time of soil sampling.
- Birds shot down (especially fledglings, assumed to be fed by the parents with feed collected at the immediate vicinity of the nests) and small vertebrates trapped. As pesticide residues were detected in the reference site, additional research was done on the supplementary reference site, located clearly outside the control zone initially demarcated.
- In addition to measurements of pollutants, inventories of bird and small mammal populations were taken.

4. RESULTS

The results reported below refer only to DDTR.

4.1. Bioaccumulation in Plants

A study of the literature allows us to forecast a certain degree of bioaccumulation in plants, according to the usual mechanisms, but not to forecast the relation between soil concentrations and concentrations in plants.

Variable quantities of pesticides were found in the samples of plants collected on the site. Low quantities of DDTR were found in berries of *Vaccinium* sp. (a few mg · kg^{-1}). No residue was found in other plants.

4.2. Bioaccumulation in Terrestrial Invertebrates

The BCF values found in the literature for organochlorine insecticides are highly variable (from 0.1 to 14). The model used here provides values ranging from 0.09 to 3.0 according to the f_{oc}. The value ultimately retained is fairly low (0.25), because of the relatively high organic matter content of soil (12% on average).

Correlations were established between soil concentrations and concentrations in the worms, measured in the bioassays. The BCF values found by these methods tend to be lower than the preceding ones (0.13).

Invertebrates collected on the site were generally less contaminated than the leaf litter in which they lived, with high variations according to the zoological group (gastropods, earthworms, millipedes, isopods, coleopteran *Carabidae*).

4.3. Impact on Terrestrial Invertebrates

Certain data in the literature suggest that the concentrations present in the soil on the site can be toxic for invertebrates. However, other data show that, even taking into account the possibility of accumulating high concentrations of pesticides, there is no manifestation of toxicity phenomena.

The bioassays are much clearer on this point and have clearly shown that the most polluted soils are toxic to earthworms.

A study of the spatial distribution of 'sentinel worms' showed a structuring of toxic zones, with a relative heterogeneity. Inventories of epigeal and hypogeal fauna indicated that, even in the most polluted soils, toxic to earthworms in bioassay conditions, the epigeal communities do not seem to be modified. On the contrary, the hypogeal fauna (for example, earthworms, *Carabus* larvae) are degenerated, in comparison to those on the least polluted sites and reference sites.

4.4. Bioaccumulation in Birds and Small Mammals

A BCF of 2.2 was calculated by the authors for DDTR from the data of Mahoney (1974) for the white-throated sparrow (*Zonotrichia albicollis*). Different exposure scenarios were compared, according to whether the birds get 10%, 50%, or 100% of their food ration from four types of contaminated feed: from invertebrates of the epigeal zone of soil (Fig. A4.1), from soil invertebrates of the highly polluted or moderately polluted sites, and from berries. The concentrations in these different sources were either measured or estimated by modelling. The results obviously varied greatly with the scenario followed. For example, the exclusive consumption of fruits led to very low estimated concentrations in birds (0.04 mg · kg^{-1}; Fig. A4.1 A), while consumption of soil invertebrates from the most contaminated zone, for only 10% of the feed ration, seriously contaminated the birds (31.7 mg · kg^{-1}; Fig. A4.1 B). Analogous models were used to study bioaccumulation in small mammals.

Field studies show that the birds contain DDTR, with median values (0.5 mg · kg^{-1}) corresponding to those obtained on the assumption that the birds have a slightly polluted feed. The highest measured value (7.7 mg · kg^{-1}; Fig. A4.1, dotted line) corresponds to a bird whose feed ration is composed of 50% earthworms from the moderately polluted site and 10% earthworms from the highly polluted site. The results are extremely variable. For catbirds (*Dumetella carolinensis*), the authors found no statistically significant differences between the polluted sites (3.67 ± 3.22 mg · kg^{-1} of DDTR) and reference sites (0.21 ± 0.17 mg · kg^{-1}), despite a strong apparent difference. However, the difference is significant when the polluted site is compared with the supplementary reference site, where the rate of contamination is 0.08 ± 0.06 mg · kg^{-1}. Higher concentrations are found in the small mammals, but the sample size must be taken into account: 2 shrews, 6 white-footed mice (*Peromyscus leucopus*), and various rodents of an unspecified number.

4.5. Toxic Effects on Birds and Small Mammals

The approach of predicting mortality is based on the hazard index defined in section 3.1. The literature does not give specific values for small songbirds. The data are those found for other groups. The authors use the lowest NOEL found in the literature for DDTR (10 mg · kg^{-1}). The highest value (350 mg · kg^{-1}) was defined as one-tenth of the highest LC$_{50}$ found in the literature (which corresponds to a safety factor of 10 to convert

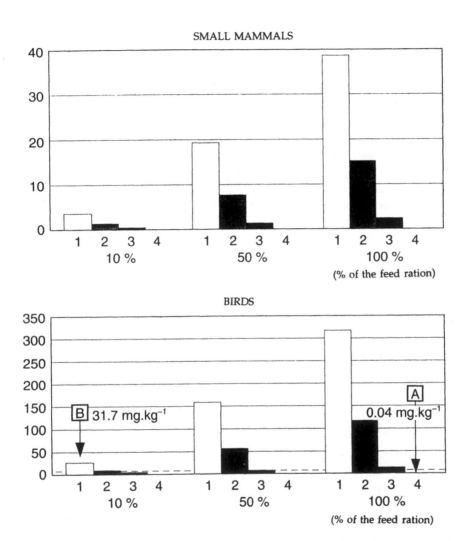

Fig. A4.1. Risks resulting from exposure of birds and small mammals to DDTR (DDT residues and metabolites of DDT) present in the soil of the Baird and McGuire site, according to three possible scenarios (Menzie et al., 1992; modified with the permission of SETAC, 1997). The bars represent the rate of contamination predicted for a bird or small mammal consuming (1) invertebrates from the epigeal zone, (2) earthworms from the highly polluted site, (3) earthworms from the moderately polluted site, and (4) berries, for 10%, 50%, and 100% of their food ration, the remainder being made up of uncontaminated feed. In calculating the predictable concentrations, the rates of contamination of invertebrates and berries were calculated from measured values, while the rates of contamination of earthworms were modelled. For birds, the horizontal line corresponds to 7.7 mg · kg^{-1} (measured value). For explanation of A and B, see text.]

LOAEL into NOEL). The published values (2 to 100 mg · kg^{-1} of DDTR) were used to estimate the effects on reproduction, but these values are very approximate. The D$_{feed}$ values used to calculate the hazard index are calculated according to three exposure scenarios (three feed regimes):

• A worst-case approach (scenario A), based on the consumption of invertebrates from the most contaminated zone (130 mg · kg^{-1}).

• A medium- or worst-case approach (scenarios B or C) based on a backcalculation of concentrations in the feed from concentrations measured in the birds and a BCF value of 2.2. With this method, the highest value measured (7.7 mg · kg^{-1}) corresponds to a concentration in the feed of 3.5 mg · kg^{-1} (scenario B), and the medium value (3.3 mg · kg^{-1}) to a concentration of 1.5 mg · kg^{-1} (scenario C).

The results, reported in Fig. A4.2, indicate a potential danger of mortality and effects on reproduction with scenario A. Scenarios B and C conclude an absence of risk of mortality, but a certain risk for reproduction, high for scenario B, low for scenario C. It is noted that the risk for reproduction is higher than the risk of mortality (the hazard indices are higher).

Field inventories revealed the presence of numerous species of birds and small mammals, in the polluted as well as the non-polluted zone. However, amphibians and reptiles were not found in the polluted zone. It was not possible to obtain quantitative data, for example, on the number of couples or the number of birds per nest, which would have provided precise measurements of successful reproduction.

5. CONCLUSIONS

The conclusions of this study are particularly interesting with regard to the value of different approaches used to evaluate ecological risk.

In the first place, the predictive methods have a very high degree of uncertainty, particularly when one attempts to evaluate the significance of transfers in the trophic chain. The 'earthworm' model is the best, with a fairly good correlation of results with those of bioassays, but the model for birds and small mammals is not as reliable, because the exposure is much more difficult to evaluate (for example, birds can find their feed outside the polluted zone or feed on plants that are more or less contaminated). The results may be compromised by errors that may reach several orders of magnitude. The authors conclude unambiguously that the predictive models currently developed are insufficient to obtain a correct estimate of exposure and effects on birds and small mammals. Samples of birds and small mammals are easy enough to procure for measurements of concentrations of pollutants and various indicators of their state of health, and they provide much more information.

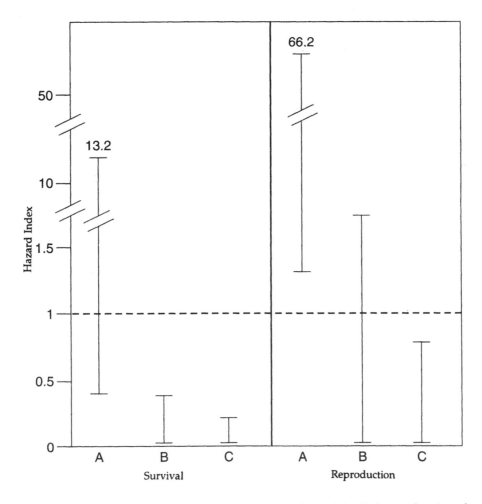

Fig. A4.2. Hazard indices (final point: survival or reproduction) for birds as a function of their feed regime (modified according to Menzie et al., 1992; with the permission of SETAC, 1997). For the different scenarios, see text.

Bioassays and the use of sentinel animals caged on the site (earthworms) give good information on the potential toxicity of soils and the potential for bioaccumulation, particularly if one combines the results of two types of approach, for example to account for differences between epigeal and endogeal communities. Earthworms caged in different locations on the site also have the advantage of providing a detailed map of exposure and toxic effects.

While recognizing that they have not succeeded in attaining the objectives stated at the beginning of the study, because of the difficulty of obtaining data on the effects on reproduction and at the community level, Menzie and team estimate that the effects observed at the individual level can indicate a possible risk at the ecosystem level. Overall, they recommend a combination of approaches. The conservatism of their conclusion is remarkable and shows clearly the degree of uncertainty that exists about the ecological effects of polluted sites, particularly the impact on terrestrial systems, even when the pollution is due to high doses of products as well studied as DDT and its metabolites. On this subject, Forbes and Forbes (1994) observe that a regulatory ban (as has been in effect for DDT and organochlorine insecticides for more than 20 years in most industrialized countries) is not enough to ensure that the problems of pollution caused by persistent products will disappear.

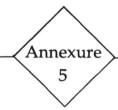

Annexure
5

Data Search

The acquisition of documented data is indispensable to any health or ecological risk evaluation. Bibliographic and factual data are compiled in databases that are highly variable in their nature, specificity, and medium of storage (published papers, on-line, CD-ROM). The accessibility of these data is a crucial point. Access to data may be limited deliberately (as when the data are the property of a particular group) or because of lack of information about their location or the material difficulty of finding and extracting them. Lack of accessibility leads to a considerable loss of information and a risk of redundant efforts in environmental management.

The objective of this annexure is not to list these databases exhaustively, but to cite those that are most accessible and useful in a risk evaluation. On the general problem of information in toxicology, see Wexler (1990).

1. BIBLIOGRAPHIC DATABASES

These databases compile data of international literature (articles, government reports, seminar proceedings, books, and book chapters). The descriptions comprise the bibliographic information and often an abstract and key words. Whether they are multidisciplinary or sector-based, none of these databases is exhaustive. The databases cited here are accessible on CD-ROM or on-line by a server via Transpac. The advent and widespread use of the Internet now allows access to a considerable quantity of data provided by research organizations, government agencies, or even small research teams, but these data must be validated.

CA Search: produced by *Chemical Abstracts Service* (Columbus, Ohio, USA); > 11 million references; since 1967. Compiles world literature on chemistry (a section is devoted to toxicology).

BIOSIS Preview: produced by BIOSIS (Philadelphia, Pennsylvania, USA); > 8.3 million references; since 1969. Compiles biological and biomedical literature, including toxicology.

PASCAL: produced by INST (Institut de l'Information Scientifique et Technique, CNRS, Vandoeuvre-les-Nancy, France); > 8.5 million references; since 1973. Multidisciplinary, good coverage of French literature.

Life Sciences Collection: produced by *Cambridge Scientific Abstracts* (Bethesda, Maryland, USA); > 2 million references; since 1978. Covers ecology, toxicology, health risk, and ecological risk.

Enviroline: produced by *Environment Abstracts* (Bethesda, Maryland, USA); > 200,000 references; since 1971. Covers all aspects of the environment, including toxicology, safety, and pollution.

Pollution Abstracts: produced by *Cambridge Scientific Abstracts* (Bethesda, Maryland, USA); since 1970. Covers all pollution issues.

Toxline: produced by the US National Library of Medicine (Bethesda, Maryland, USA); since 1971. Made up of a number of thematic subfiles, concerns the toxic or undesirable effects of chemical and physical agents and medicines on biological systems.

NTIS: produced by the National Technical Information Service (Springfield, Virginia, USA); > 2 million references; since 1964. Compiles research reports or studies financed by the US government and federal and local agencies, and some reports financed by Japan, Great Britain, Germany, and France (CNRS).

Medline and **Excerpta Medica** are essentially medical databases, useful only for health risk.

2. FACTUAL DATABASES

Factual databases provide data (numerical or verbal) that are directly usable (e.g., LC_{50}, LD_{50}, daily admissible doses, description of symptoms, lesions, toxic interactions, regulatory values). Unlike bibliographic databases, some are directly accessible in the documentary databases, but most of the information is compiled in the publications themselves or in bibliographic reviews (for example, *Reviews in Environmental Contamination and Toxicology*).

The most numerous and the most accessible databases in the field of ecological risk cover *chemical substances* (common name, synonyms, CAS registration number, molecular formula, physico-chemical properties, production) and their *toxicity* (acute and chronic toxicity, mutagenicity, teratogenicity, carcinogenicity, effects on reproduction, etc.). Among the most important are the following:

Chemtox Online: produced by Resource Consultants (Brentwood, Tennessee, USA). 10,300 descriptions in 1995. Covers substances that have been or are potentially the subject of regulation or legislation. The database contains substances identified and regulated by the EPA and various American laws, such as the Resource Conservation and Recovery Act, the

Clean Air Act, the Clean Water Act, the Toxic Substances Control Act, the Superfund Amendment and Reauthorization Act, and the Federal Insecticide, Fungicide and Rodenticide Act.

The data are extracted from lists published by the US Federal Register, the code of federal regulations, US government publications, journals, reports, and manufacturers' safety files.

RTECS (Registry of Toxic Effects of Chemical Substances): produced by the National Institute of Occupational Safety and Health (Cincinnati, Ohio, USA); > 100,000 compounds; since 1971. Covers the basic toxicology of medicines, agrochemical products, food additives, chemical wastes, and their products of reaction. Also contains maximum exposure doses recommended by the NIOSH and US federal regulations as well as data not published by the EPA.

MERCK INDEX ONLINE: produced by Merck (Rahway, New Jersey, USA); from the 19th century to 1993. Corresponds to the printed version of Merck Index 11th edition.

Environmental Chemical Data and Information Network (ECDIN): produced by the Commission of European Communities, Brussels, Belgium; 65,000 references in 1992. Covers the effects of chemical products on the environment, the actual and potential risks linked to the use of chemical products, ecological and economic impact.

Environmental Fate: produced by the EPA;. 8000 references in 1992. Describes the occurrence and behaviour of chemical substances in the environment. The products listed are manufactured in quantities greater than 450 tonnes per year (450 chemical substances in 1992).

In the field of agrochemical products (pesticides, fungicides, and herbicides), several databases are easily searched, including the following:

European Directory of Agrochemical Products: produced by the Royal society of Chemistry (Cambridge, UK); 25,000 compounds manufactured and marketed in Europe. Formulations can be found with the percentage of active ingredient for each country, application deadlines, toxicity, use, and use limitations.

The Agrochemicals Handbook: produced by the Royal Society of Chemistry (Cambridge, UK); since 1983. Indicates the common name, synonyms, CAS registration number, formula, molecular weight, physicochemical properties, mode of action, activity, and toxicity of active ingredient used in the manufacture of agrochemical products.

AGRITOX: produced by the INRA (Institut National de la Recherche Agronomique, France). Indicates for each active ingredient the physicochemical properties, results of toxicity and ecotoxicity tests, occurrence and behaviour in soil, and-administrative arrangements (notably the maximum residue limits).

Databases concerning professional toxicology are not cited here. They hold little interest for the study of ecological risk.

There are currently many factual databases in toxicology, primarily of US origin. We have seen that they cover the physico-chemical properties of products and the results of toxicity and ecotoxicity tests. Other information may be useful in ecological risk evaluation, such as data on emissions of *toxic products* (nature, volume, or location). The EPA and the National Library of Medicine have compiled inventories of chemical substances manufactured in or imported into the United States, for example, *the TSCA Chemical Substances Inventory*, the *Toxics Release Inventory* (TRI), or the *Hazardous Substances DataBank* (HSDB). However, these databanks do not list direct measurements of exposure.

It is relatively easy to find factual data on the environment itself (such as climatic characteristics or nature of soils). It is harder to find factual data on the *occurrence of products, exposure,* and *toxic effects on populations.* With regard to the occurrence of products, STORET (Storage and Retrieval of Water Quality Data) is an example of a database compiling information on water quality (surface water and phreatic layer) in the United States. This database is designed for health risk, but such information can be used for ecological risk.

Databases of measurements of *in situ* indicators are much fewer in number. They concern essentially health risk, for example, *the National Human Adipose Tissue Survey* (NHATS), which aims to 'detect and quantify the prevalence of toxic organic compounds in the overall population'. To our knowledge, there are no centralized files on exposure data or toxic effects for wild fauna.

The role of an *exposure database* would be to help estimate individual exposure, trace the distribution of exposure levels in the population, measure total exposure and the proportion of each avenue of exposure in total exposure, follow the occurrence of a product in the environment, record the biological and environmental samples, etc. (Burke et al., 1992; Goldman et al., 1992). The present databases are very imperfect. The chief difficulties are the following:
- reliability of data: what degree of confidence can one have in the data?
- their heterogeneity: how do we treat data from various sources?
- their representativity: of which populations are the data a sample?

Very often, the data lack information on several points:
- the *definition* of variables: for example, chemical names, identification of diseases;
- the experimental *protocols*, notably their conformity to good laboratory practices;
- the identification of the population sampled (notably location, ethnic characteristics);
- the *periods* of sampling.

The *representativity* of data is difficult to establish. If they are not *all* counted, the sample is biased. Unfortunately, many data remain inaccessible because they are not well disseminated (as with internal reports). Sometimes, the data are not published for more subtle reasons: some groups tend to stress publication of positive data (that is, data showing toxic effects of a product) rather than negative data, while other groups do the opposite.

References

Abdul-Rida AMM, Bouché M (1994). A method to assess chemical biorisks in terrestrial ecosystem. In *Ecotoxicology of soil organisms*. Donker MH, Eusackers H, Heimbach F (eds). Lewis Publishers, Boca Raton, pp. 388–394.

Académie des Sciences (1995). Biodiversité et environnement. Rep. no. 33, TEC & DOC, Lavoisier, Paris, 96 pp.

Addiscott TM, Wagenet RJ (1985). Concepts of soluble leaching in soil: a review of modeling approaches. *J. Soil. Sci.* 36:411–424.

AFNOR (1992). Détermination de l'indice biologique global normalisé (IBGN. No NF-T90350), AFNOR, Paris, 9 pp.

AFNOR (1994). Qualité des sols. Collection environnement, Lavoisier, Paris, 250 pp.

AFNOR (1995). Essais écotoxicologiques. Collection environnement. Lavoisier, Paris, 336 pp.

Aldenberg T, Slob W (1993). Confidence limits for hazardous concentrations based on logistically distributed NOEC toxicity data. *Ecotox. Environ. Saf.* 25:48–63.

Anderson J, Kieser S, Bean R, Riley R, Thomas B (1981). Toxicity of chemically dispersed oil to shrimp exposed to constant and decreasing concentrations in a flowing system. In *Proceedings of the 1981 Oil Spill Conference*. American Petroleum Institute, Washington, DC, pp. 69–75.

Ankely GT, Schubauer-Berigan MK, Dierkes JR (1991). Predicting the toxicity of bulk sediments to aquatic organisms with aqueous test fractions: pore water vs elutriate. *Environ. Toxicol. Chem.* 10:1359–1366.

Asher SC, Lloyd KM, Mackay D, Paterson S, Roberts JR (1985). A critical examination of environmental modeling—Modeling the fate of chlorobenzenes using the persistence and fugacity models. National Research Council of Canada.

ASTM (1987). *Standard guide for conducting a terrestrial soil–core microcosm test*. 1991 Annual Book of ASTM Standards. vol. 11.04.E 1197–87:819–831, ASTM, Philadelphia.

Athey LA, Thomas JM, Miller WE, Ward JQ (1989). Evaluation of bioasays for designing sediment cleanup strategies at a wood treatment site. *Environ. Toxicol. Chem.* 8:223–230.

Bacci E, Renzoni A, Gaggi C, Calamari D, Franchi A, Vighi M, Severi A (1990a). Models, field studies, laboratory experiments: an integrated approach to evaluate the environmental fate of atrazine (s-triazine herbicide). In *Agricultural ecology and environment: proceedings of an international symposium*. Paoletti MG, Stinner BR, Lorenzoni GG (eds.). Padova, Elsevier, Amsterdam.

Bacci E, Calamari D, Gaggi G, Vighi M (1990b). Bioconcentration of chemical vapors in plant leaves: Experimental measures and correlation. *Environ. Sci. Technol.* 24:885–889.

Baes CF (1982). Environmental transport and monitoring. Prediction of radionuclide Kd values from soil–plant concentration ratios. *Trans. Amer. Nucl. Soc.* 41:53.

Baes CF, Sharp RD, Sjoreen AL, Shor RW (1984), A review and analysis of parameters of assessing transport of environmentally released radionuclides through agriculture. ORNL-5786. Oak Ridge National Laboratory. Oak Ridge. Tennessee.

Bailey HC, Liu DHW, Javitz HA (1984). Time/toxicity relationships in short-term static, dynamic and plug-flow bioassays. In *Aquatic Toxicology and Hazard Assessment*. Bahner RC, Hansen, DJ (eds.). ASTM STP 891, ASTM, Philadelphia, pp. 193–212.

Baize D, Jabiol B (1995). Guide pour la description des sols. Editions de l'INRA, Coll. Techniques et pratiques, Versailles, 375pp.

Baker SS (1989). Assessment strategies and approaches. In *Ecological assessment of hazardous waste sites: A field and laboratory reference*. Warren-Hicks W, Parkhurst BR, Baker SS (eds.). EPA/600/3–89/013, Corvallis.

Barbault R (1992). *Écologie des peuplements. Structure, dynamique et évolution*, 2nd ed. Masson, Paris, 273 pp.

Barbault R (1993). *Écologie générale. Structure et fonctionnement de la biosphère*. Masson, Paris, 269 pp.

Barber MC, Suarez LA, Lassiter RR (1988). Modeling bioconcentration of nonpolar organic pollutants by fish. *Environ. Toxicol. Chem.* 7:545–558.

Barnthouse LW (1992). The role of models in ecological risk assessment: a 1990's perspective. *Environ. Toxical. Chem.* 11:1751–1760.

Barron MG (1990). Bioconcentration. *Environ. Sci. Technol.* 24:1612–1618.

Bartell SM, Gardner RH, O'Neill RV (1992), *Ecological risk estimation*. Lewis Publishers, Chelsea, 252 pp.

Bascietto J, Hinckley D, Plafkin J, Slimak M (1990). Ecotoxicity and ecological risk assessment. Regulatory applications at EPA. First part of a four-part series. *Environ. Sci. Technol.* 24:10–15.

Baughman GL, Lassiter RR (1978). Prediction of environmental pollution. In *Estimating the hazard of chemical substances to aquatic life*. Cairns J, Dickson KL, Maki AW (eds.). ASTM, STP 657, Philadelphia, pp. 35–54.

Bengtsson G, Torstensson L (1988). Soil biological variables in environmental hazard assessment. Concepts for a research program. *National Swedish Environmental Protection Board, Solna*, NEPBR, rep. 3499.

Berlan-Darqué M, Kalaora B (1989). Environnement et identité. À propos d'un projet d'enquête nationale. *SRETIE-Info* 26/27:17–19, Ministère de l'Environnement, Paris.

Blanck H, Wänberg SA, Molander S (1988). Pollution induced community tolerance—a new ecotoxicological tool. In *Functional testing of aquatic biota for estimating hazard of chemicals*. Cairns J, Pratt JR (eds.). ASTM STP 988. ASTM, Philadelphia. pp. 219–230.

Blandin P (1986). Bio-indicateurs et diagnostic des systèmes écologiques. *Bull. Ecol.* 17:215–307.

Block JC, Bauda P, Reteuna C (1989). Ecotoxicity testing using aquatic bacteria. In *Aquatic ecotoxicology: fundamental concepts and methodologies*, vol. 2. Boudou A, Ribeyre F (eds.). CRC Press, Boca Raton, pp. 187–210.

Boersma L, Lindstrom T, McFarlane C, McCoy EL (1988). Uptake of organic chemicals by plants: A theoretical model. *Soil. Sci.* 146:403–417.

Boersma L, McFarlane C, Lindstrom T (1991). Mathematical model of plant uptake and translocation of organic chemicals: application to experiments. *J. Environ. Qual.* 20:137–146.

Boese BL, Lee H, Sprecht DT, Randall RC, Winsor MH (1990). Comparison of aqueous and solid phase uptake for hexachlorobenzene in the tellinid clam *Macoma nasuta* (Conrad): a mass balance approach. *Environ. Toxicol. Chem.* 9:221–231.

Bogen KT, Spear RC (1987). Integrating uncertainty and interindividual variability in environmental risk assessment. *Risk Anal.* 7:427–435.

Bouché M (1988). Earthworm toxicological tests, hazard assessment and biomonitoring. A methodological approach. In *Earthworms in waste and environmental management*. Edwards CA, Neuhauser EF (eds.). SPB Publications, The Hague, Netherlands, pp. 315–320.

Bretthauer EW (1993). EPA's approach to environmental research in the 90s. *Environ. Toxicol. Chem.* 12:1331–1333.

Briggs GG (1981). Theoretical and experimental relationships between soil adsorption, octanol–water partition coefficients, water solubilities, bioconcentration factors, and the parachor, *J. Agric. Food Chem.* 29:1050–1059.

Briggs GG, Bromilow RH, Evans AA (1982). Relationship between lipophilicity and root uptake and translocation of non-ionized chemicals by barely. *Pestic. Sci.* 13:495–504.

Briggs GG, Bromilow RH, Evans AA, Williams M (1983). Relationship between lipophilicity and the distribution of non-ionized chemicals by barley shoots following uptake by the roots. *Pestic. Sci.* 14:492–500.

Brown DS, Flagg EW (1981). Empirical prediction of organic pollutant in natural sediments. *J. Environ. Qual.* 10:382–386.

Buck WB (1979). Animals as monitors of environmental quality. *Vet. Hum. Toxicol.* 21:277–284.

Burger J, Gochfeld M (1992). Temporal scales in ecological risk assessment. *Arch. Environ. Contam. Toxicol.* 23:484–488.

Burke T, Anderson H, Beach N, Colome S, Drew RT, Firestone M, Hauchman FS, Miller TO, Wagener DK, Zeise L, Tran N (1992). Role of exposure databases in risk management. *Arch. Environ. Health* 47:421–429.

Burmaster DE, Anderson PD (1994). Principles of good practice for the uses of Monte-Carlo techniques in human health and ecological risk assessments. *Risk Anal.* 14:477–481.

Cairns J (1988). Should regulatory criteria and standards be based on multispecies evidence. *Environ, Prof.* 10: 157–165.

Cairns J (1992). The threshold problem in ecotoxicology. *Ecotoxicology* 1:3–16.

Cairns J (1993). Communication and status: the dilemma of an environmental scientist. *Specul. Sci. Technol.* 16:163–170.

Cairns J, MacCormick PV (1991). The use of community- and ecosystem-level end points in environmental hazard assessment: a scientific and regulatory evaluation. *Environ. Audit* 2:239–248.

Carins J, Niederlehner BR (1993). Ecological function and resilience: neglected criteria for environmental impact assessment and ecological risk analysis. *Environ. Prof.* 15:116–124.

Cairns J, Dickson KL, Maki AW (1978). *Estimating the hazard of chemical substances to aquatic life.* STP 657, ASTM, Philadelphia.

Cairns J, MacCormick PV, Belanger SE (1992). Ecotoxicological testing: small is reliable. *J.Environ. Pathol. Toxicol. Oncol.* 11:247–263.

Cairns J, MacCormick PV, Belanger SE (1993a). Prospects for the continued development of environmentally-realistic toxicity tests using microorganisms. *J. Environ. Sci.* 5:265–268.

Cairns J, MacCormick PV, Niederlehner BR (1993b). A proposed framework for developing indicators of ecosystem health. *Hydrobiologia* 263:1–44.

Calamari D, Vighi M (1992). Role of evaluative models to assess exposure to pesticides. In *Methods to assess adverse effects of pesticides on non-target organisms.* Tardiff RG (ed.). SCOPE 49, John Wiley and Sons, Chichester, pp. 119–132.

Calder WA, Braun EJ (1993). Scaling of osmotic regulation in mammals and birds. *Regulatory Integrative Comp. Physiol.* 13:R601–R606.

Calow P (1993). Overview with observations on risk assessment and management. In *Handbook of ecotoxicology.* Calow P (ed.). Blackwell Scientific Publications, Oxford.

Calow P, ed. (1993). *Handbook of Ecotoxicology,* vol. 1. Blackwell Scientific Publications, Oxford, 478 p.

Calow P, ed. (1994). *Handbook of Ecotoxicology,* vol. 2. Blackwell Scientific Publications, Oxford, 416 p.

Chapman PM (1989). Current approaches to developing sediment quality criteria. *Environ. Toxicol. Chem.* 8:589–599.

Chapman PM (1991). Environmental quality criteria: what type should we be developing? *Environ. Sci. Technol.* 25:1353–1359.

Chiou CT, Peters LJ, Freed VH (1979). A physical concept of soil–water equilibria for non-ionic organic compounds, *Science* 106:831–832.

Chiou CT, Porter PE, Schemmding DW (1983). Partition equilibria of non-ionic organic compounds between soil organic matter and water. *Environ. Sci. Technol.* 17:227–231.

Clark T, Clark K, Paterson S, Mackay D, Norstrom RJ (1988). Wildlife monitoring, modeling and fugacity, *Environ, Sci. Technol.* 22:120–127.

Clark TP, Norstrom RJ, Fox GA, Won HT (1987). Dynamics of organochlorine compounds in herring gulls (*Larus argentatus*). II. A two-compartment model and data for ten compounds. *Environ. Toxicol. Chem.* 6:547–559.

Conn, WD, Rich RC (1992). Communicating about ecosystem risks. In *Predicting ecosystem risk*, Advances in Modern Environmental Toxicology, vol. 30. Cairns J, Niederlehner BR, Orvos DR (eds.). Princeton Scientific Publishing, Princeton, pp. 1–8.

Connell DW (1989). *Bioaccumulation of xenobiotic compounds*. CRC Press, Boca Raton.

Costanza R, Norton B, Haskell B, eds. (1992). *Ecosystem health: new goals for environmental management*. Island, Washington, DC.

Costanza R, Principe PP (1995). Methods for economic and sociological considerations in ecological risk assessment. In *Methods to Assess Effects of Chemicals on Ecosystems*. Linthurst RA, Bourdeau P, Tardiff RG (eds.). SCOPE 53, Wiley and Sons, Chichester, pp. 395–406.

Courtemanch DL, Davies SP (1987). A coefficient of community loss to assess detrimental change in aquatic community. *Water. Res.* 21:217–222.

Covello VT, Merkhofer MW (1993). *Risk assessment methods*. Plenum Press, New York, 318 pp.

Croft BA, Morse JG (1979). Research advances in pesticide resistance to natural enemies. *Entomophaga* 24:3–11.

CSA (1986). *Canadian Standard Association. Guidelines for calculating derived release limits for radio-active material in airborne and liquid effluents for normal operations of nuclear facilities.* CSA Standards (draft).

Dab W (1994). Évaluation des risques sanitaires liés à l'environnement et décision en santé publique. École Nationale de Santé Publique, Département 'Environnement Santé'.

Davis DE, Winstead RL (1980). Estimating the numbers of wildlife populations. In *Wildlife management techniques manual*, 4th ed. Schemnitz SD (ed.). The Wildlife Society, Washington, DC.

DeAngelis DL, Barnthouse LW, Van Winkle W, Otto RG (1990). A critical appraisal of population approaches in assessing fish community health. *J. Gt Lakes Res.* 16:576–590.

DeAngelis DL, Gross LJ (1992). *Individual-based models and approaches in ecology*. Chapman and Hall, New York.

DECHEMA (1995). Bioassays for soils. In *Fourth report of the Interdisciplinary DECHEMA Committee 'Environmental biotechnology—soil.'* Kreysa G, Wiesner J (eds.). Frankfurt-on-Main, 45 pp.

DeGraeve GM, Cooney JD, Marsh BH, Pollock TL, Reichenback NG (1992). Variability in the performance of the 7-d *Ceriodaphnia dubia* survival and reproduction test: an intra- and interlaboratory study. *Environ. Toxicol. Chem.* 11:851–866.

DeSnoo GR, Canters KJ, DeJong FMW, Cuperus R (1994). Integral hazard assessment of side effects of pesticides in the Netherlands—A proposal. *Environ. Toxicol. Chem.* 13:1331–1340.

Devillers J, Domine D, Karcher W (1995). Estimating *n*-octanol/water partition coefficients from the autocorrelation method. *SAR QSAR Environ. Res.* 3:301–306.

de Zwart D, Slooff W (1983). The Microtox as an alternative assay in the acute toxicity assessment of water pollutants. *Aquat. Toxicol.* 4:129–138.

DGM (1990). *Premises for risk management: risk limits in the context of environmental policy*. Directorate-General for Environmental Protection at the Ministry of Housing. Physical Planning and Environment, Leidschendam, Netherlands.

DHHS (1986). Determining risks to health. Federal Policy and Practice, Task Force on Health Risk Assessment. Department of Health and Human Services, Dover, USA.

Dickson KL, Maki AW, Brungs WA (1987). *Fate and effects of sediment-bound chemicals in aquatic systems*. Pergamon Press, New York.

Dijkshoorn W, Lampe JEM, van Broekhoeven LW (1981). Influence of soil pH on heavy metals in ryegrass from sludge amended soil. *Plant and Soil* 61:277.

Dijkshoorn W, Lampe JEM, van Broekhoeven LW (1983a). Effect of soil pH and ammonium and nitrate treatments on heavy metals in ryegrass from sludge amended soil. *Neth. J. Agric. Soil.* 31:181.

Dijkshoorn W, Lampe JEM, van Broekhoeven LW (1983 b). The effect of soil pH and chemical form of nitrogen fertilizer on heavy metals in ryegrass. *Fertil. Res.* 4:63.

DiToro DM (1985). A particlë interaction model of reversible organic chemical sorption. *Chemosphere,* 14:1503–1508.

Dobson S (1993). Why different regulatory decisions when the scientific information base is similar?—Environmental risk assessment. *Regul. Toxicol. Pharmacol.* 17:333–345.

Donaldson WT (1992). The role of property–reactivity relationships in meeting the EPA's needs for environmental fate constants. *Environ. Toxicol. Chem.* 11:887–891.

Donkin P (1994). Quantitative structure–activity relationships. In *Handbook of Ecotoxicology,* Vol. 2. Calow P (ed.). Blackwell Scientific Publications, pp. 321–347.

Doty CB, Travis CC (1990). Is EPA's national priorities list correct? *Environ. Sci. Technol.* 24:1778–1780.

Doull J (1996). Keynote address, *Drug Metab. Rev.* 28:1–7.

Doust JL, Schmidt M, Doust LL (1994). Biological assessment of aquatic pollution: a review, with emphasis on plants as biomonitors. *Biol. Rev.* 69:147–186.

Dreicer M, Hakanson TE, White GC, Whicker FW (1984). Rainsplash as a mechanism of soil contamination of plant surfaces. *Health Phys.* 46:177–187.

Dybing E (1995). EUROTOX–SOT debate: In assessing the safety of a chemical mixture, the mixture itself should be tested. *EUROTOX Newsletter* 18:7–9.

Eberhart LL, Thomas JM (1991). Designing environmental field studies. *Ecol. Monogr.* 61:53–73.

ECAO (1987). Recommendations for and documentation of biological values for use in risk assessment, EPA/600-6-87-008. Environmental Criteria and Assessment Office, Cincinnati, Ohio.

ECETOC (1992). *Estimating environmental concentrations of chemicals using fate and exposure models.* Tech. Rep. no. 50, European Chemical Industry, Ecology and Toxicology Center, Brussels.

ECETOC (1993). Environmental hazard assessment of substances. Tech Rep. no. 51, European Chemical Industry, Ecology and Toxicology Center, Brussels, 92 pp.

ECETOC (1994). *Évaluation du risque présenté par les contaminants chimiques du sol.* Tech. Rep. no. 40. (Trans. from English by Chimie et Ecologie). European Chemical Industry, Ecology and Toxicology Center, Brussels, 107 pp.

Eijsackers HJP (1989). *The Netherlands Integrated Soil Research Program: plan and realization of the research program.* Program Office for Integrated Soil Research, Wageningen.

Emans HJB, Plassche EJVD, Canton JH, Okkerman PC, Sparenburg PM (1993). Validation of some extrapolation methods used for effect assessment. *Environ. Toxicol. Chem.* 12:2139–2154.

Emlen JM (1989). Terrestrial population models for ecological risk assessment: a state-of-the-art review. *Environ. Toxicol. Chem.* 8:831–842.

Enfield GC, Carsel RF, Cohen SZ, Phon T, Walters DM (1982). Approximating pollutant transport to groundwater. *Ground Water* 20:711–727.

EPA (1982). Appendix A—Uncontrolled hazardous waste site scoring system: A user's manual. Feb. Reg. 37:31219–31243.

EPA (1985). Superfund Public Health Assessment Manual (draft). Washington, DC.

EPA (1986). Guidelines for estimating exposures. Fed. Reg. 51:34042–34054.

EPA (1987a). Biological criteria for the protection of aquatic life, vol. II. Users manual for biological risk assessment of Ohio surface waters. EPA, Columbus, Ohio.

EPA (1987b). Biological criteria for the protection of aquatic life, vol. III. Standardized biological field sampling and laboratory methods for assessing fish and macroinvertebrates communities. EPA, Columbus, Ohio.

EPA (1988). Superfund exposure assessment manual. EPA 540/1-88-001. Office of Remedial Response, Washington, DC.

EPA (1989a). Risk assessment guidance for Superfund, vol. I. Human health evaluation manual (Part A). Interim final. EPA/540/1-89/002. Office of Emergency and Remedial Response, Washington, DC.

EPA (1989b). Supplemental risk assessment guidance for the Superfund program. Region 1, Risk Assessment Workgroup, Boston.

EPA (1989c). Risk assessment guidance for Superfund, vol. II. Environmental evaluation manual, Interim final. EPA/540/1-89/001, Office of Emergency and Remedial Response, Washington, DC.

EPA (1990). Hazard Ranking System: final rule. Fed. Reg. 55:51532–51667.

Erickson PA (1994). *A practical guide to environmental impact assessment.* Academic Press, San Diego, 266 pp.

Evans HE, Dillon PJ (1995). Methods to evaluate whole aquatic and terrestrial systems. In *Methods to assess effects of chemicals on ecosystems.* Linthurst RA, Bourdeau P, Tardiff RG (eds.). SCOPE 53, Wiley and Sons, Chichester, pp. 337–354.

Everts JW, Eys Y, Ruys M, Pijnenburg J, Visser H, Luttik R (1993). Biomagnification and environmental quality criteria: a physiological approach. ICES *J. Mar. Sci.* 50:333–335.

Finkel A (1989). Is risk assessment really too conservative: revising the revisionists. *Columbia J. Environ. Law* 14:427–467.

Fletcher JS, Johnson FL, McFarlane JC (1990). Influence of greenhouse versus field testing and taxonomic differences on plant sensitivity to chemical treatment. *Environ. Toxicol. Chem.* 9:769–776.

Forbes VE, Forbes TL (1994). *Ecotoxicology in theory and practice.* Chapman and Hall, New York, 220 pp. French translation (1997). 'Ecotoxicologie: théorie et applications, Editions de l'INRA, Versailles.

Ford J (1989). The effects of chemical stress on aquatic species composition and community structure. In *Ecotoxicology: problems and approaches.* Levin SA, Harwell MA, Kelly JR, Kimball KD (eds.). Springer-Verlag, Heidelberg, pp. 99–144.

Fordham CL, Reagan DG (1991). Pathways analysis method for estimating water and sediment criteria at hazardous waste sites. *Environ. Toxicol. Chem.* 10:949–960.

Forsyth DJ, Peterle TJ (1984). Species and age differences in accumulation of Cl-DDT by voles and shrews in the field. *Environ. Pollut.* 33:327–340.

Fouchécourt MO, Rivière JL (1996). Activities of liver and lung cytochrome P-450-dependent monooxygenases and antioxidant enzymes in laboratory and wild Norway rats exposed to reference and contaminated soils. *Arch. Environ. Contam. Toxicol.* 30:513–522.

Fowle JR, Sexton K (1992). EPA priorities for biologic markers research in environmental health. *Environ. Health Perspect.* 98:235–24.

Franke C, Studinger G, Berger G, Böhling S, Bruckmann U, Cohors-Fresenborg D, Jöhncke U (1994). The assessment of bioaccumulation. *Chemosphere* 29:1501–1514.

Freundenburg WR, Rursch JA (1994). The risks of 'Putting the numbers in context'. *Risk Anal.* 14:949–958.

Garten CT, Trabalka JR (1983). Evaluation of models for predicting terrestrial food chain behavior of xenobiotics. *Environ. Sci. Technol.* 17:590–595.

Gaudet C, EVS Environment Consultants, Environment and Social Systems Analysts (1994). Cadre de travail pour l'évaluation du risque écologique que présentent les lieux contaminés situés au Canada: études et recommandations. Environnement Canada, Série Scientifique, Study no. 199, 110 pp.

Gelber RD, Lavin PT, Mehta CR, Schoenfeld DA (1985). Statistical analysis. In *Fundamentals of aquatic toxicology: Methods and applications.* Rand GM, Petrocelli SR (eds.). Hemisphere, Washington, DC.

Gerstl Z, Mingelgrin U (1984). Sorption of organic subsances by soils and sediments. *J. Environ. Sci. Health* B19:297–312.

Geyer M, Sheehan P, Kolzias D, Freitag D, Korte F (1982). Prediction of ecotoxicological behaviour of chemicals: Relationship between physico-chemical properties and bioaccumulation of organic chemical on the mussels, *Mytilus edulis. Chemosphere* 11:1121.

Geyer HJ, Scheunert I, Rapp K, Gebefügl I, Steinberg C, Kettrup A (1993). The relevance of fat content in toxicity of lipophilic chemicals to terrestrial animals with special reference to dieldrin and 2,3,7,8-tetrachlorodibenzo(p)dioxin. *Ecotoxicol. Environ. Saf.* 26:45–60.

Giesy JP, Hoke RA (1989). Freshwater sediment toxicity bio-assessment: rationale for species selection and test design. *J. Gt Lakes Res.* 15:539–569.

Gilbertson M, Kubiak T, Ludwig J, Fox G (1991). Great Lakes embryo mortality, edema, and deformities syndrome (GLEMEDS) in colonial fish-eating birds. Similarity to chickedema disease. *J. Toxicol. Environ. Health* 33:455–520.

Gillett JW (1989). The role of terrestrial microcosmes and mesocosms in ecotoxicological research. In *Ecotoxicology: problems and approaches.* Levin SA, Harwell MA, Kell JR, Kimball KD (eds.). Springer-Verlag, Heidelberg. pp. 367–410.

Gobas FAPC, McNeil EJ, Lovett-Doust L, Haffner GD (1991). Bioconcentration of chlorinated aromatic hydrocarbons in aquatic macrophytes. *Environ. Sci. Technol.* 25:924–929.

Gobas FAPC, Z'Graggen MN, Zhang X (1995). Time response of the lake Ontario eco-system to virtual elimination of PCBs. *Environ. Sci. Technol.* 29:2038–2046.

Goldman LR, Gomez M, Greenfiled S, Hall I, Hulka BS, Kaye WE, Lybarger JA, McKenzie DH, Murphy RS, Wellington DG, Woodruff T (1992). Use of exposure databases for status and trends analysis. *Arch. Environ. Health* 47:430–438.

Gordon GE (1988). Receptor models. *Environ. Sci. Technol.* 22:1132–1142.

Gray JS (1980). The measurement of effects of pollutants on benthic communities. *Rapp. P-v réun. Cons. Int. Explor. Mer.* 179:188–193.

Gray RH (1993). Case study: applying total quality environmental management to Hanford's public outreach and involvement efforts. *Fed. Facil. Environ. J. (Summer)* 231–243.

Gray RH, Becker CD (1993). Environmental clean-up: the challenge at the Hanford site, Washington, USA. *Environ. Mgmt.* 17:461–475.

Green RH (1965). Estimation of tolerance over an indefinite time period. *Ecology* 46:887.

Greene JC, Miller WE, Debacon M, Long MA, Bartels CL (1988a). Use of *Selenastrum capricornutum* to assess the toxicity potential of surface and ground water contamination caused by chromium waste. *Environ. Toxicol. Chem.* 7:35–39.

Greene JC, Bartels CL, Warren-Hicks WJ, Parkhurst BR, Linder GL, Peterson SA, Miller WE (1988b). Protocols for short-term toxicity screening of hazardous waste sites. US. EPA, 3-88-029, 102 pp.

Greig–Smith PW (1992). A European perspective on ecological risk assessment, illustrated by pesticide registration procedures in the United Kingdom. *Environ. Toxicol. Chem.* 11:1673–1689.

Hair JD (1980). Measurement of ecological diversity. In *Wildlife management techniques manual,* 4th ed. Schemnitz SD (ed.). The Wildlife Society, Washington, DC.

Hansch C, Maloney PP, Fujita T, Muir RM (1962). Correlation of biological activity of phenoxyacetic acids with Hammett substituent constants and partition coefficients. *Nature* 194:178–180.

Harris HJ, Wenger RB, Harris VA, Devault DS (1994). A method for assessing environmental risk: a case study of Green Bay, Lake Michigan, USA. *Environ. Mgmt.* 18:295–306.

Harwell MA, Cooper W, Flaak R (1992). Prioritizing ecological and human welfare risks from environmental stresses. *Environ. Mgmt.* 16:451–464.

Hassett JJ, Bawart WL, Griffin RA (1983). Correlation of compound properties with sorption characteristics of non-polar compounds by soils and sediments: concepts and limitations. In *Environment and solid waste.* Francis CW, Auerbach SI (eds.). Butterworth, Boston, pp. 161–178.

Heasley JE, Lauenroth WK, Dodd JL (1981). Systems analysis of potential air pollution impacts on grassland ecosystems. In *Energy and ecological modeling.* Mitsch WJ, Bosserman RW, Klopatek JM (ed.). Elsevier, New York, pp. 347–359.

Hellawell JM (1986). *Biological indicators of freshwater pollution and environmental management.* Elsevier, Amsterdam, 546 pp.

Hermens JLM (1989). Quantitative structure-activity relationships of environmental pollutants. In *Handbook of environmental chemistry,* vol. 2. Hutzinger O (ed.). Springer-Verlag, Berlin, pp. 112–162.

Hersh CM, Crumpton WG (1987). Determination of growth rate depression of some green algae by atrazine. *Bull. Environ. Contam. Toxicol.* 39:1041–1048.

Hetrick DM, MacDowell-Boyer LM (1983). User's manual for TOXSCREEN: A multimedia screening-level program for assessing the potential fate of chemicals released to the environment. ORNL/TM-8570. Oak Ridge. National Laboratory, Oak Ridge, Tennessee.

Hoekstra JA (1991). Estimation of the LC_{50}, a review. *Environmetrics* 2:139–152.

Hope BK (1995). A review of models for estimating terrestrial ecological receptors exposure to chemical contaminants. *Chemosphere* 30:2267–2287.

Hülster A, Müller JF, Marschner H (1994), Soil–plant transfer of polychlorinated dibenzo-p-dioxins and dibenzofurans to vegetables of the cucumber family (Cucurbitaceae). *Environ. Sci. Technol.* 28:1110–1115.

Hunsaker CT, Graham RL, Barnthouse LW, Gardner RH, O'Neill RV, Suter GW (1990). Assessing ecological risk on a regional scale. *Environ. Mgmt.* 14:325–332.

Hunsaker C, Graham R, Turner RS, Ringold PL, Holdren GR, Strickland TC (1993). A national critical loads framework for atmospheric deposition effects assessment: II. Defining assessment end points, indicators, and functional subregions. *Environ. Mgmt.* 17:335–341.

12C2 (1994). INSA/INRA/CRIDEAU/CNRS: *Investigation sur les différentes approaches de la définition et de la qualification des sites et sols pollués.* Rep. 93-503, Association RECORD, Lyon.

Ireland FA, Judy BM, Lower WR, Thomas MW, Krause GF, Asfaw A, Sutton WW (1991). Characterization of eight soil types using the *Selenastrum capricornutum* bioassay. In *Plants for toxicity assessment*, vol. 2. Gorsuch JW, Lower WR, Wang W, Lewis MA (eds.). ASTM STP 1114, pp. 217–219, ASTM, Philadelphia.

ISO (1991a). Soil quality—effects of pollutants on earthworms (*Eisenia fetida*). Method for the determination of acute toxicity using artificial soil substrate, IOS, Paris, TC190/SC4/WG2 N20.

ISO (1991b). Soil determination of the effect of chemical substances on the reproduction of earthworms. ISO, Paris, TC190/SC4/WG2 N27.

Jepson PC (1993). Insects, spiders and mites. In *Handbook of ecotoxicology*. Calow P (ed.). Blackwell Scientific Publications. Oxford.

Johnson AR (1988). Diagnostic variables as predictors of ecological risk. *Environ. Mgmt.* 12:515–523.

Jouany JM, Vaillant M, Blarez B, Cabridenc R, Ducloux R, Schmitt S (1982). Approach to hazard assessment by a qualitative system based on interaction between variables. *Proceedings of the Symposium. 'Chemicals in the Environment. Chemicals Testing and Hazard Ranking,'* Copenhagen. 18–20 Oct.

Juday C, Schloemer CL, Livingston C (1938). Effect of fertilisation on plankton production and fish growth in a Wisconsin lake. *Prog. Fish Culturist* 40:24–27.

Jury WA, Spencer WF, Farmer WJ (1983). Behavior assessment model for trace organics in soil. *J. Environ. Qual.* 12:558–564.

Jury WA, Ghrodati M (1989). Overview of organic chemical environmental fate and transport modeling approaches. In *Reactions and movement of organic chemicals in soil*. Sawhney BL, Brown K (eds.). Madison, USA.

Kabata-Pendias A, Pendias H (1984). *Trace elements in soils and plants* . CRC Press, Boca Raton.

Kaputska LA (1989). Vegetation assessment. In *Ecological assessment of hazardous waste sites: A field and laboratory reference*. Warren-Hicks W, Parkhurst BR, Baker SS (eds.). EPA/600/3-89/013. Corvallis, USA.

Kaputska LA, Reporter M (1993). Terrestrial primary producers. In *Handbook of ecotoxicology*, vol. 1. Calow P (ed.). Blackwell Scientific Publications, Oxford, pp. 278–298.

Kara AH, Hayton WL (1984). Pharmacokinetic model for the uptake and disposition of di-2-ethyl phthalate in sheepshead minnow *Cyprinodon variegatus*. *Aquat. Toxicol* 5:181–195.

Karichof SW (1981). Semi-empirical estimation of sorption of hydrophobic pollutants on natural sediments and soil. *Chemosphere* 10:833–846.

Karr JR (1981). Assessment of biotic integrity using fish community. *Fisheries* 6:21–27.

Karr JR (1991). Biological integrity: A long-neglected aspect of water resource management. *Ecol. Appl.* 1:66–84.

Karr JR, Fausch KD, Angermeier PL, Yant PR, Schloser IJ (1986). Assessing biological integrity in running waters: a method and its rationale. Special Publication 5. Illinois Natural History Survey. Champaign.

Keddy C, Greene JC, Bonnell MA (1994). Examen des biotests effectués sur des organismes entiers pour l'évaluation de la qualité des sols, des sédiments et des eaux douces au Canada. Environnement Canada. Study no. 198. Série scientifique. Ottawa.

Keddy C, Greene JC, Bonnell MA (1995). Review of whole-organisms bio-assays: soil, freshwater sediment and freshwater assessment in Canada. *Ecotoxicol. Environ. Sci.* 30:221–251.

Kenaga EE (1982). Predictability of chronic toxicity from acute toxicity of chemicals in fish and aquatic invertebrates, *Environ. Toxicol. Chem.* 1:347–358.

Kepner CH, Tregoe BB (1965). The rational manager, a systematic approach to problem solving and decision making. Kepner-Tregoe Inc., Princeton, USA.

Klein W, Koerdel W, Klein AW, Kuhnen-Clausen D, Weiss M (1988). Systematic approach for environmental hazard ranking of new chemicals. *Chemosphere* 7:1445–1462.

Könneman WH (1981). Quantitative structure–activity relationships in fish toxicity studies. 1. Relationship for 50 industrial pollutants. *Toxicology* 19:209–221.

Kooijman SALM (1981). Parametric analyses of mortality rates in bioassays. *Water Res.* 15:107–119.

Kooijman SALM (1987). A safety factor for LC_{50} values allowing for differences in sensitivity among species. *Water Res.* 21:269–276.

Labinenic PA, Dzombak DA, Siegrist RL (1996). SoilRisk: Risk assessment model for organic contaminants in soil *J. Environ. Eng.* 122:338–398.

Lambolez L, Vasseur P, Ferard JF, Gisbert T (1994). The environmental risks of industrial waste disposal: an experimental approach including acute and chronic toxicity studies. *Ecotoxicol. Environ. Saf.* 28:317–328.

Landis WG, Yu MH (1995). *Introduction to environmental toxicology.* Lewis Publishers, Boca Raton, 328 pp.

Landres PB, Verner J, Thomas JW (1988). Ecological uses of vertebrate indicator species: a critique. *Conserv. Biol.* 2:316–328.

Landrum PF, Lee H, Lydy MJ (1992). Toxicokinetics in aquatic systems. Model comparisons and use in hazard assessment. *Environ. Toxicol. Chem.* 11: 1709–1725.

LaPoint TW, Fairchild JF (1989). Aquatic surveys. In *Ecological assessment of hazardous waste sites: A field and laboratory reference.* Warren-Hicks W, Parkhurst BR, Baker SS (eds.). EPA/600/3-89/013, Corvallis.

LaPoint TW, Perry JA (1989). Use of experimental ecosystems in regulatory decision making. *Environ. Mgmt.* 13:539–544.

Legator MS, Strawn SF (1993). Public health policies regarding hazardous waste sites and cigarette smoking: An argument by analogy. *Environ. Health Perspect.* 101:8–11.

Leon CD, Van Gestel (1994). *Selection of a set of laboratory ecotoxicity tests for the effects assessment of chemicals in terrestrial ecosystems.* Rep. no. D94009, Department of Ecology and Ecotoxicology, Free University of Amsterdam, Amsterdam, 135 pp.

Lewis MA (1993). Freshwater primary producers. In *Handbook of ecotoxicology,* vol. 1. Calow P (ed.). Blackwell Scientific Publications, Oxford, pp. 28–50.

Linder G, Richmond ME (1990). Feed aversion in small mammals as a potential source of hazard reduction for environmental chemicals: agrochemical case studies. *Environ. Toxicol. Chem.* 9:95–105.

Linthurst RA, Bourdeau P, Tardiff RG, eds. (1995). *Methods to assess effects of chemicals on ecosystems.* SCOPE 53, Wiley and Sons, Chichester, 416 pp.

Lipton J, Galbraith H, Burger J, Wartenberg D (1993). A paradigm for ecological risk assessment. *Environ. Mgmt.* 17:1–5.

Logan DT, Wilson HT (1995). An ecological risk assessment method for species exposed to contaminant mixtures. *Environ Toxicol. Chem.* 14:351–359.

Lokke H, Van Gestel CAM (1993). Development, improvement and standardisation of test systems for assessing sublethal effects of chemicals on fauna in the soil ecosystem. *National Environmental Research Institute*, Silkeborg, 41 pp.

Lowrance R, Vellidis G (1995). A conceptual model for assessing ecological risk to water quality function of bottomland hardwood forests. *Environ. Mgmt.* 19:239–258.

MacBee K, Bickham JW (1990). Mammals as bioindicators of environmental toxicity. *Curr. Mammal* 2:37–88.

MacCarty LS, Mackay D, Smith AD, Ozburn GW, Dixon DG (1993). Interpreting aquatic toxicity QSARs: the significance of toxicant body residues at the pharmacologic endpoint. *Sci. Tot. Environ.* 109/110:515–525.

MacCarthy JF, Shugart LR (1990). *Biomarkers of environmental contamination.* Lewis, Boca Raton, 457 pp.

MacIntosh DL, Suter GW, Hoffman FO (1994). Use of probabilistic exposure models in ecological risk assessments of contaminated sites. *Risk Anal.* 14405–419.

Mackay D (1991). *Multimedia environmental models: The fugacity approach.* Lewis Publishers and Chelsea.

Mackay D (1994). Fate models. In *Handbook of Ecotoxicology*, Vol. 2. Calow P (ed.). Blackwell Scientific Publications, Oxford. pp. 348–367.

Mackay D, Paterson S (1991). Evaluating the multimedia fate of organic chemicals: a level III fugacity model. *Environ. Sci. Technol.* 25:427–436.

Mackay D, Paterson S, Shiu WY (1992). Generic models for evaluating the regional fate of chemicals. *Chemosphere* 24:695–717.

MacKone TE (1993). The precision of QSAR methods for estimating intermedia transfer factors in exposure assessments. *SAR and QSAR in Environ. Res.* 1:41–51.

MacKone TE, Layton DW (1986). Screening the potential risk of toxic substances using a multimedia compartment model: Estimation of model exposure. *Regul. Toxicol. Pharmacol.* 6:359.

MacKone TE, Ryan PB (1989). Human exposures to chemicals through food chains: an uncertainty analysis. *Environ. Sci. Technol.* 23:1154–1163.

Madhum YA, Freed VH, Young JL (1986). Binding of ionic and neutral herbicides by soil humic acid. *Soil Sci. Soc. Am. J.* 50:319–322.

Mahoney J (1974). Residue accumulation in white-throated sparrows fed DDT for five and 11 weeks, *Bull. Environ. Contam. Toxicol.* 12:677–681.

Maki AW, Duthie JR (1978). Summary of proposed procedures for the evaluation of aquatic hazards. In *Estimating the hazard of chemical substances to aquatic life.* Cairns J, Dickson KL, Maki AW (eds.). ASTM STP 657, ASTM, Philadelphia.

Markwell RD, Connell DW, Gabric AJ (1989). Bioaccumulation of lipophilic compounds from sediments by oilgochaetes. *Water Res.* 23:1443–1450.

Mayer FL, Mayer KS, Ellersieck MR (1986). Relationship of survival to other end-points in chronic toxicity tests with fish. *Environ. Toxicol. Chem.* 5:737–748.

Mayfield CI (1993). Microbial systems. In *Handbook of ecotoxicology*, Vol. 1. Calow P (ed.). Blackwell Scientific Publications, Oxford, pp. 9–27.

Menzie CA, Burmaster DE, Freshman JS, Callahan CA (1992). Assessment of methods for estimating ecological risk in the terrestrial component: A case study at the Baird and McGuire Superfund site in Holbrook, Massachusetts. *Environ. Toxicol. Chem.* 11:245–260.

Merkhofer MW (1987). *Decision science and social risk management: a comparative evaluation of cost–benefit analysis, descision analysis, and other formal decision aiding approaches.* D Reidel, Dordrecht, Netherlands.

Metcalf R (1971). A model ecosystem for the evaluation of pesticide biodegradability and ecological magnification. *Outlook Agric.* 7:55–69.

Mill T (1993). Environmental chemistry. In *Ecological risk assessment.* Suter GW (ed.). Lewis Publishers, Chelsea, 538 pp.

Mill T, Walton BT (1987). How reliable are data-base data? *Environ. Toxicol. Chem.* 6:161–162.

Mineau P, ed. (1991). *Cholinesterase-inhibiting insecticides. Their impact on wildlife and the environment.* Elsevier, Amsterdam, 348 pp.

Mineau P, Collins BT, Baril A (1996). On the use of scaling factors to improve inter-species extrapolation of acute toxicity in birds. *Regul. Toxicol. Pharmacol.* 24:24–29.

Moreale A, van Bladel R (1981). Adsorption de 13 herbicides et insecticides par le sol. Relation solubilité-réactivité. *Revue Agric.* 34: 939–952.

Morgan E, Knacker T (1994). The role of laboratory terrestrial model ecosystems in the testing of potentially harmful substances. *Ecotoxicology* 3: 213–233.

Moriarty F (1988). *Ecotoxicology: The study of pollutants in ecosystems*, 2nd ed. Academic Press, New York.

Morris P, Therivel R (1995). *Methods of environmental impact assessment.* The Natural and Built Environment Series 2, UCL Press. London, 378 pp.

Mount DI (1992). What evidence of ecosystem risk is necessary to influence regulatory and industrial decisions? In *Predicting Ecosystem Risk, Advances in Modern Environmental Toxicology*, vol. 30. Cairns J, Niederlehner BR, Orvos DR (eds.). Princeton Scientific Publishing, Princeton, pp. 31–37.

Munawar M, Munawar F, Mayfield CI, McCarthy LH (1989). Probing ecosystem health: a multi-disciplinary and multi-trophic assay strategy. *Hydrobiologia* 188/189: 93–116.

National Research Council (1983). *Risk assessment in the Federal Government: Managing the Process.* National Academy Press, Washington, DC.

National Research Council (1991). *Animals as sentinels of environmental health hazards.* National Academy Press, Washington, DC, 160 pp.

National Research Council (1994). *Science and judgment in risk assessment.* National Academy Press, Washington, DC, 649 pp.

Nendza M (1991). QSARs in bioconcentration: Validity assessment of logPow/logBCF correlations. In *Bioaccumulation in aquatic systems.* Nagel R, Loskill R (eds.). VCH. Weinheim.

Newman JR, Schreiber RK (1984). Animals as indicators of ecosystem responses to air emissions. *Environ. Mgmt.* 8:309–324.

Nilsson R, Tasheva M, Jaeger B (1993). Why different regulatory decisions when the scientific information base in similar? Human risk assessment. *Regul. Toxicol. Pharmacol.* 17:292–332.

Norton SB, Rodier DJ, Gentile JH, Van der Schalie WH, Wood WP, Slimak MW (1992). A framework for ecological risk assessment at the EPA. *Environ. Toxicol. Chem.* 11:1663–1672.

O'Brien DJ, Kaneene JB, Poppenga RH (1993). The use of mammals as sentinels for human exposure to toxic contaminants in the environment. *Environ. Health. Perspect.* 99:351–368.

OCDE (1981). Guidelines for testing of chemicals No. 305E. Bio-accumulation: essai dynamique chez le poisson. OCDE, Paris.

OCDE (1984). Ligne directrice de l'OCDE pour less essais de produits chimiques: plantes terrestres, essai de croissance.

OCDE (1989). Compendium of environmental exposure assessment methods for chemicals. Monograph no. 27, OCDE, Paris.

OCDE (1992a). The OECD principles of Good Laboratory Practice. Monograph no. 45, OCDE, Paris.

OCDE (1992b). Report of the OECD Workshop on Quantitative Structure–Activity Relationships (QSARs). Monograph no. 58, OCDE, Paris.

OCDE (1993a). Application of structure-activity relationships to the estimation of properties important in exposure assessment. Monograph no. 67, OCDE, Paris, 62 pp.

OCDE (1993b). Structure–activity relationships for biodegradation. Monograph no. 68, OCDE, Paris, 103 pp.

OCDE (1993c). Report of the OECD workshop on the application of simple models for environmental assessment. Monograph no. 69, OCDE, Paris, 143 pp.

OCDE (1995). Report of the OECD workshop on environmental hazard/risk assessment. Monograph no. 105, OCDE, Paris, 96 pp.

OEPP (1994). Système de décision pour l'évaluation des effets non intentionnels des produits phytosanitaires sur l'environnement—Vertébrés terrestres. Decision-making scheme for the environmental risk assessment of plant protection products—Terrestrial vertebrates. *Bulletin EPPO/OEPP* 24:37–87.

Olie K, van den Berg M, Hutzinger O (1983). Formation and fate of PCDD and PCDF from combustion processes. *Chemosphere* 12:627.

Osborne LL, Davies RW, Linton KJ (1980). Use of hierarchical diversity indices in lotic community analysis. *J. Appl. Ecol.* 17:567–580.

Pascoe GA, DalSoglio JA (1994a). Planning and implementation of a comprehensive ecological risk assessment at the Milltown reservoir–Clark Fork river Superfund site, Montana. *Environ. Toxicol. Chem.* 13:1943–1956.

Pascoe GA, DalSoglio JA (1994b). Ecological assessment for the wetlands at Milltown reservoir, Missoula, Montana: Characterization of emergent and upland habitats. *Environ. Toxicol. Chem.* 13:1957–1970.

Pascoe GA, DalSoglio JA (1994c). Characterization of ecological risks at the Milltown reservoir–Clark Fork river sediments Superfund site, Montana. *Environ. Toxicol. Chem.* 13:2043–2058.

Pastorok RA, Peek DC, Sampson JR, Jacobson MA (1994). Ecological risk assessment for river sediments contaminated by creosote. *Environ. Toxicol. Chem.* 13:1929–1941.

Paterson S, Mackay D, Tam D, Shiu WY (1990). Uptake of organic chemicals by plants: a review of processes, correlations and models. *Chemosphere* 21:297–331.

Patin SA (1982). *Pollution and the biological resources of the oceans.* Butterworth Scientific, London.

Payne JF, Fancey LL, Rahimtula AD, Porter EL (1987). Review and perspective on the use of mixed-function oxygenase enzymes in biological monitoring. *Comp. Biochem. Physiol.* 86C:233–245.

Peakall DB, Tucker RK (1985). Extrapolation from single species studies to populations, communities, and ecosystems. In *Methods for estimating risk of chemical injury: human and non-human biota and ecosystems.* Vouk VB, Butler GC, Hoel DG, Peakall DB (eds.). SCOPE 26, Wiley and Sons, New York, pp. 611–636.

Peakall DB, Shugart LR (1993). *Biomarkers: research and application in the assessment of environmental health.* NATO ASI Series, Serie H: Cell Biology, vol. 68. Springer-Verlag, 119 pp.

Persoone G, Janseen CR (1993). Freshwater invertebrate toxicity tests. In *Handbook of Ecotoxicology*, vol. 1. Calow P (ed.). Blackwell Scientific Publications, Oxford, pp. 51–65.

Peters HP (1993). In search for opportunities to raise 'environmental risk literacy'. *Toxicol. Environ. Chem.* 40:289–300.

Peterson BJ, Fry B (1987). Stable isotopes in ecosystem studies. *Ann. Rev. Ecol. Syst.* 18:293–320.

Petts J (1994). Effective waste management: understanding and dealing with public concerns. *Waste Mgmt. Res.* 12:207–222.

Phillips DJH (1993). Bioaccumulation. In *Handbook of Ecotoxicology*, vol. 1. Calow P (ed.). Blackwell Scientific Publications, Oxford, pp. 378–396.

Pontasch KW, Niederlehner BR, Cairns J (1989). Comparisons of single-species, microcosm and field responses to a complex effluent. *Environ. Toxicol. Chem.* 8:521–532.

Ramade F (1992). *Précis d'écotoxicologie.* Masson, Paris, 300 pp.

Ramade F (1993). *Dictionnaire encyclopédique des sciences de l'environnement.* Ediscience, Paris, 822 pp.

Rand GM, Petrocelli SR (1985). *Fundamentals of aquatic toxicology.* Hemisphere, New York.

Rapport DJ (1989). What constitutes ecosystem health? *Perspect. Biol. Med.* 33:120–132.

Reades DW, Gorber DM (1986). A site–specific approach for the development of soil clean–up criteria for trace organics. Shell's Oakville Refinery, Golder Associates/SENES Consultants Ltd.

Reteuna C, Vasseur P, Cabridenc R (1989). Performances of three bacterial assays in toxicity assessment. *Hydrobiologia* 188/189:149–153.

Reynoldson TB, Day KE (1993). Freshwater sediments. In *Handbook of Ecotoxicology*. vol. 1. Calow P (ed.). Blackwell Scientific Publications, Oxford, pp. 83–100.

Riederer M (1990). Estimating partitioning and transport of organic chemicals in the foliage/atmosphere system: Discussion of a fugacity-based model. *Environ. Sci. Technol.* 24:829–837.

Rivière JL, Fouchécourt MO, Walker CH (in press). Biomarqueurs d'exposition des animaux terrestres aux polluants. In *Utilisation de biomarqueurs en écotoxicologie.* Lagadic L, Caquet T, Ramade R (eds.). Masson, Paris.

Roberts JR (1992). Exposure assessment models: accuracy and validity. In *Methods to assess adverse effects of pesticides on non-target organisms*. Tardiff RG (ed.). SCOPE 49, John Wiley and Sons, Chichester, pp. 133–148.

Roberts JR, Mitchell MS, Boddington MJ, Ridgeway JM (1981). A screen for the relative persistence of lipophilic organic chemicals in aquatic ecosystems—An analysis of the role of a simple computer model in screening, Part I. National Research Council of Canada, Rep. no. NRCC 18570, Ottawa.

Rodricks JV (1994). Risk assessment, the environment, and public health. *Environ. Health Perspect.* 102:258–264.

Romijn CFA, Luttik R, Meet D, Sloof W, Canton JH (1993). Presentation of a general algorithm to include effect assessment on secondary poisoning in the derivation of environmental quality criteria. Part 1. Aquatic food chains. *Ecotoxicol. Environ. Saf.* 26:61–85.

Requeplo P (1996). Entre savoir et décision, l'expertise scientifique. INRA-Editions, Versailles, 112 pp.

Rundgren S, Andersson R, Brinkman L, Byman J, Gustafsson K, Johnsson I, Torstensson L (1989). Soil biological variables in environmental hazard assessment: organisation and research program. National Swedish Environmental Protection Board, Solna, Sweden. NSEPB Rep. no. 3603.

Ryan JA, Bell RM, Davidson JM, O'Connor GA (1988). Plant uptake of non-ionic organic chemicals from soils. *Chemosphere* 17:2299.

Saarikowski J, Viluksela M (1982). Relation between physicochemical properties of phenols and their toxicity and accumulation in fish. *Ecotoxicol. Environ. Saf.* 6:501–512.

Sabljic A, Piver WT (1992). Quantitative modeling of environmental fate and impact of commercial chemicals. *Environ. Toxicol. Chem.* 11:961–972.

Sauerbeck D (1988). Der Transfer von Schwermetallen in die Pflanze. In *Beurteilung von Schwermetallkontaminationen im Boden*. DECHEMA, Francfort.

Schaeffer DJ, Herricks EE, Kerster HW (1988). Ecosystem health 1. measuring ecosystem health. *Environ. Mgmt.* 12:445–455.

Scientist (1994). Long-awaited risk assessment commission finally ready to convene 8(9):2 May 1994.

Seed J, Brown RP, Olin SS, Foran JA (1995). Chemical mixtures: current risk assessment methodolopies and future directions. *Regul. Toxicol. Pharmacol.* 22:76–94.

Senes Consultants (1989). Contaminated Soil Clean-up in Canada, vol. 5, Development of the AERIS Model. Final Report prepared for the Decommissioning Steering Committee.

SETAC (1991a). Workshop on aquatic microcosms for ecological assessment of pesticides, Wintergreen, 56 pp.

SETAC (1991b). Guidance document on testing procedures for pesticides in freshwater mesocosms, Huntingdon, 46 pp.

Sheehan PJ (1995). Assessment of ecological impacts on a regional scale. In *Methods to assess effects of chemicals on ecosystems*. Linthurst RA, Bourdeau P, Tardiff RG (eds.). SCOPE 53, Wiley and Sons, Chichester, pp. 306–336.

Sheehan PJ, Miller GC, Butler C, Bourdeau P, eds. (1984). *Effects of pollutants at the ecosystem level*. SCOPE 22, John Wiley & Sons, New York, 460 pp.

Shrader-Frechette KS (1994). Science, environmental risk assessment, and the frame problem. *Bioscience* 44:548–551.

Shu HD, Paustenback D, Murray L, Marple L, Brunck B, Dei Rossi D, Teitelbaum P (1988). Bioavailability of soil-bound TCDD: oral bioavailability in the rat. *Fund. Appl. Toxicol.* 10:648–654.

Silbergeld EK (1993). Risk assessment: the perspective and experience of U.S. environmentalists. *Environ. Health Perspect.* 101:100–104.

Simonich SL, Hites RA (1994). Vegetation–atmosphere partitioning of polycyclic aromatic hydrocarbons. *Environ. Sci. Technol.* 28:939–943.

Sipes IG, Gandolfi AJ (1991). Biotransformation of toxicants. In *Casarett and Doull's Toxicology: The Basic Science of Toxicants*, 4th ed. Amdur MO, Doull J, Klaassen CD (eds.). Pergamon Press, New York.

Slooff W (1983). Benthic macroinvertebrates and water quality assessment: Some toxicological considerations. *Aquat. Toxicol* 4:113–128.

Slooff W, Canton JH, Hermens JLM (1983). Comparison of the susceptibility of 22 freshwater species to 15 chemical compounds. I. (Sub)-acute tests. *Aquat. Toxicol.* 4:73–82.

Slooff W, Van Oers JAM, de Zwart D (1986). Margins of uncertainty in ecotoxicological hazard assessment. *Environ. Toxicol. Chem.* 5:841–852.

Smith ED, Barnthouse LW (1987). *User's manual for the defense priority model.* Oak Ridge National Laboratory, ORNL-6411, Oak Ridge, Tennessee.

Smith EP, Cairns J (1993). Extrapolation methods for setting ecological standards for water quality: Statistical and ecological concerns. *Ecotoxicology* 2:203–219.

Smith EP, Pontasch KW, Cairns J (1989). Community similarity and the analysis of multi-species environmental data: a unified statistical approach. *Water Res.* 24:507–514.

Smith RH (1993). Terrestrial mammals. In *Handbook of Ecotoxicology*, vol. 1. Calow P (ed.). Blackwell Scientific Publications, Oxford, pp. 339–352.

Solbè JF de LG (1993). Freshwater fish. In *Handbook of Ecotoxicology*, vol. 1. Calow P (ed.). Blackwell Scientific Publications, Oxford, pp. 66–82.

Spencer WF, Farmer W, Cliath MM (1973). Pesticide volatilization. *Residue Rev.* 49:1–27.

Stafford EA, Tacon AG (1988). Use of earthworms as a food for rainbow trout. In *Earthworms in waste and environmental management*. Edwards CA, Neuhauser EF (eds.). SBP Academic Publishing. The Hague, pp. 193–208.

Stallones L, Nuckols JR, Berry JK (1992). Surveillance around hazardous waste sites: geographic information systems and reproductive outcomes. *Environ. Res.* 59:81–92.

Steinberg CEW, Geyer HJ, Kettrup AAF (1994). Evaluation of xenobiotic effects by ecological techniques. *Chemosphere* 28:357–374.

Stephan CE, Mount DI, Hansen DJ, Gentile JH, Chapman GA, Brungs WA (1985). *Guidelines for deriving numerical national water quality criteria for the protection of aquatic organisms and their uses.* Duluth, US Environmental Research Laboratories.

Suntio LR, Shiu WY, Mackay D, Seiber JN, Glotfelty D (1988). Critical review of Henry's law constant for pesticides. *Rev. Environ. Contam. Toxicol.* 103:1–59.

Suter GW (1987). Species interactions. In *Methods for assessing the effects of mixtures of chemicals.* Vouk VB, Butler GC, Upton AC, Parke DV, Asher SC (eds.). Scope 30, Wiley & Sons, Chichester, pp. 745–758.

Suter GW (1990a). Environmental risk assessment/environmental hazard assessment: similarities and differences. In *Aquatic toxicology and risk assessment*, vol. 13. Landis WG, Van Der Schalie WH (eds.). ASTM, Philadelphia.

Suter GW (1990b). Uncertainty in environmental risk assessment. In *Acting under uncertainty: multidisciplinary conceptions.* von Furstenberg GM (ed.). Kluwer Academic Publishers, USA, pp. 203–230.

Suter GW (1993a). *Ecological risk assessment.* Lewis Publishers, Chelsea, 538 pp.

Suter GW (1993b). A critique of ecosystem health concepts and indexes. *Environ. Toxicol. Chem.* 12:1533–1539.

Suter GW (1994). Endpoints of interest at different levels of biological organization. In *Ecological toxicity testing. Scale, complexity and relevance.* Cairns J, Niederlehner BR (eds.). Lewis Publishers, Boca Raton, pp. 35–48.

Suter GW, Loar JM (1992). Weighing the ecological risk of hazardous waste sites: the Oak Ridge case. *Environ. Sci. Technol.* 26:432–438.

Suter GW, Rosem AE, Linder E (1986). Analysis of extrapolation error. In *User's manual for ecological risk assessment.* Barnthouse LW, Suter GW (eds.). ORNL-6251, Oak Ridge National Laboratory, Oak Ridge, pp. 49–81.

Suter GW, Vaughan DS, Gardner RH (1983). Risk assessment by analysis of extrapolation error, a demonstration for effects of pollutants on fish. *Environ. Toxicol. Chem.* 2:369–378.

Talmage SS, Walton BT (1991). Small mammals as monitors of environmental contaminants. *Rev. Environ. Contam Toxicol.* 119:47–109.

Taylor GE Jr, Hanson PJ, Baldocchi DD (1988). Pollutant deposition to individual leaves and plant canopies: Sites of regulation and relationship to injury. In *Assessment of crop loss from air pollutants*. Heck WW, Taylor OC, Tingey DT (eds.). Elsevier Publishers, New York.

Thomann RV (1989). Bioaccumulation model of organic chemical distribution in aquatic food chains. *Environ. Sci. Technol.* 23: 699–707.

Thomas JM, Skalski JR, Cline JR, McShane MC, Simpson JC, Miller WE, Peterson SA, Callahan SA, Greene JC (1986). Characterization of chemical waste site contamination and determination of its extent using bioassays. *Environ. Toxicol. Chem.* 5:487–501.

Thompson M (1994). Blood, sweat and tears. *Waste Mgmt. Res.* 12:199–205.

Topp EI, Schenert A, Attar A, Korte A, Korte F (1986). Factors affecting the uptake C14-labelled organic chemicals by plant from soil. *Ecotoxicol. Environ. Saf.* 11:219.

Travis CC, Arms AD (1988). Bioconcentration of organics in beef, milk and vegetation. *Environ. Sci. Technol.* 22:271–274.

Travis CC, Hattemeyer-Frey HA (1988). Uptake of organics by aerial plant parts: a call for research. *Chemosphere* 17:227–283.

Truhaut R (1977). Écotoxicologie: objectifs, principes et perspectives. *Ecotoxicol. Environ. Saf.* 1:151–173.

Tullock G, Brady G (1992). The risk of crying wolf. In *Predicting ecosystem risk, Advances in Modern Environmental Toxicology*, vol. 30. Cairns J, Niederlehner BR, Orvos DR (eds.). Princeton Scientific Publishing, Princeton pp. 77–91.

Urban DJ, Cook NJ (1986). Hazard evaluation, standard evaluation procedure, ecological risk assessment. EPA-540/9-85-001, US Environmental Protection Agency, Washington, DC.

Van Gestel CAM, Ma WC (1988). Toxicity and bioaccumulation of chlorophenols in earthworms in relation to bioavailability in soil. *Ecotoxicol. Environ. Saf.* 15:289–297.

Van Leeuwen K (1990). Ecotoxicological effects assessment in the Netherlands. *Environ. Mgmt.* 14:779–792.

Van Straalen NM, Denneman CAJ (1989). Ecotoxicological evaluation of soil quality criteria. *Ecotoxicol. Environ. Saf.* 18:241–251.

Van Straalen NM, Van Gestel CAM (1993). Soil invertebrates and micro-organisms. In *Handbook of Ecotoxicology*. Calow P. (ed.). Blackwell Scientific Publications, Oxford.

Vasseur P, Ferrard JF, Rast C, Larbaigt D (1984). Luminescent marine bacteria in ecotoxicity screening tests of complex effluents. In *Toxicity screening procedures using bacterial systems*. Liu DL, Dutka BJ (eds.). Marcel Dekker, New York, pp. 22–36.

Verhaar HJM (1994). Predictive methods in aquatic toxicology. Thesis, University of Utrecht.

Verhoef HA, van Gestel CAM (1995). Methods to assess the effects of chemicals on soils. In *Methods to assess effects of chemicals on ecosystems*. Linthurst RA, Bourdeau P, Tardiff RG (eds.). SCOPE 53, Wiley and Sons, pp. 337–354.

Verneaux J (1976a). Application de la méthode des « indices biotiques » à l'échelle d'un réseau hydrographique: cartographie de la qualité biologique des eaux. In *La pollution des eaux continentales. Incidence sur les biocénoses aquatiques*. Pesson P (ed.). Gauthier-Villars, Paris, 265 pp.

Verneaux J (1976b). Fondements biologiques et écologiques de l'étude de la qualité des eaux continentales. Principals méthodes. In *La pollution des eaux continentales. Incidence sur les biocénoses aquatiques*. Pesson P (ed.). Gauthier-Villars, Paris, 265 pp.

Verneaux J (1995). Biodiversity in the assessment of freshwater quality. In *Methods to assess effects of chemical on ecosystems*. Linthurst RA, Bourdeau P, Tardiff RG (eds). SCOPE 53. Wiley and Sons, pp. 97–112.

Volmer J, Kordel W, Klein V (1988). A proposed method for calculating taxonomic-group-specific variances for use in ecological risk assessment. *Chemosphere* 17:1493–1500.

Vowles PD, Mantoura RFC (1987). Sediment–water partition coefficient an HPLC retention factors of aromatic hydrocarbons. *Chemosphere* 16:103–116.

Wagner C, Lokke H (1991). Estimation of ecotoxicological protection levels from NOEC toxicity data. *Water Res.* 25:1237–1242.

Walker CH (1993). Birds. In *Handbook of Ecotoxicology*. Vol. 1. Calow P (ed.). Blackwell Scientific Publications, Oxford, pp. 336–338.

Walters CJ (1986). *Adaptative Management of Renewable Resources*. MacMillan, New York.

Wang W, Freemark K (1995). The use of plants for environmental monitoring and assessment. *Ecotoxicol. Environ. Saf.* 30:289–301.

Warren-Hicks W, Parkhurst BR, Baker SS (1989). Ecological assessment of hazardous waste sites: A field and laboratory reference. EPA/600/3–89/013, Corvallis.

Weber CI, Peltier TJ, Norberg-King TJ, Horning WB, Kessler FA, Menkedick JR, Nieheisel TW, Lewis PA, Klemm DJ, Pickering QH, Robinson EL, Lazorchak J, Wymer LJ, Freyberg RW (1989). Short-term methods for estimating the chronic toxicity of effluents and receiving waters to freshwater organisms, 2nd ed. US EPA, Cincinnati, 600/4-89/001, 248 pp.

Weinstein LH, Laurence JA, Mandl RH, Wälti K (1991). Use of native and cultivated plants as bioindicators and biomonitors of pollution damage. In *Plants for toxicity assessment*. Wang W, Gorsuch JW, Lower WR (eds.). ASTM STP 1115, ASTM, Philadelphia.

Wexler P (1990). The framework of toxicology information. *Toxicology* 60:67–98.

Wheatley GO, Hardman JO (1968). Organochlorine insecticide residue in earthworms from arable soils. *J. Sci. Food Agric.* 19:219–229.

Whelan G, Strenge DL, Droppo JG, Steelman BL, Buck JW (1987). The remedial action priority action system (RAPS): Mathematical formulations. PNL-6200, Richland, Washington DC.

Wright JF, Armitage PD, Furse MT, Moss D (1989). Prediction of invertebrate communities using stream measurements. *Regulated Rivers: Research and Management* 4:147–155.

Yoshida K, Shigeoka T, Yamauchi F (1987). Multiphase non-steady state equilibrium model for evaluation of environmental fate of organic chemicals. *Toxicol. Environ. Chem.* 15:159.

Index

Milton Keynes UK
Ingram Content Group UK Ltd.
UKHW020821141024
449569UK00008B/510